Neural Aspects of Human Movement

Implications for control and coordination

The editors of this book were supported by the Foundation for Biophysics (grant number 810-403-073) and Psychon (grant number 560-259-045 and 560-259-050) which are funded by the Netherlands Organisation for Scientific Research (NWO). The editors are indebted to the Faculty of Human Movement Sciences of the Vrije Universiteit (Amsterdam), and to the Stichting Vrije Universiteitsfonds for their financial support.

C.M.C. Bakker, M.A.M. Berger, C.A.M. Doorenbosch, C.E. Peper,
M.E.T. Willems, & F.T.J.M. Zaal
Editors

Faculty of Human Movement Sciences
Vrije Universiteit, Amsterdam

Neural Aspects of Human Movement

Implications for control and coordination

SWETS & ZEITLINGER

AMSTERDAM ∎ LISSE ∎ BERWYN, PA   ∎ ACADEMIC PUBLISHING DIVISION ∎

Library of Congress Cataloging-in-Publication Data

(applied for)

Cip-gegevens Koninklijke Bibliotheek, Den Haag

Neural

Neural aspects of human movement : implications for control and coordination /
C.M.C. Bakker ... [et al.] - Amsterdam [etc.] : Swets & Zeitlinger
Uitg. n.a.v. het gelijknamige symposium, gehouden op 22 mei 1992 aan de Vrije
Universiteit Amsterdam, Faculty of Human Movement Sciences. - Met index, lit. opg.
ISBN 90-265-1322-4
NUGI 821
Trefw.: motoriek

Cover photograph: Le Penseur, Rodin
Cover design: Rob Molthoff
Cover printed in the Netherlands by Casparie, IJsselstein
Printed in the Netherlands by Offsetdrukkerij Kanters B.V., Alblasserdam

ISBN 90 265 1322 4
NUGI 821

Preface

The Graduate Institute of Human Movement (GIHM) which unifies the PhD students of the Faculty of Human Movement Sciences annually presents a symposium. At the Faculty human movement is studied in a unique interdisciplinary way. Since the boundaries between 'traditional' disciplines are crossed, the various disciplines form, at least to a certain extent, fields of interest for all Faculty members and students. Where the first symposium (1991) stressed the interdisciplinary approach to the study of human movement, the 1992 symposium directed the attention to research and theory concerning neural aspects of human movement: At May 22, 1992 the symposium *Neural aspects of human movement: Implications for control and coordination* was held. In this book the presentations as well as the main issues raised during the discussions are gathered.

Four prominent speakers shared their knowledge about the state of the art in the research on several aspects of neural processes related to movement with the audience. Following each presentation a panel of three discussants opened the discussion. During the morning session, concerning the complex neuroanatomy and the functional organisation of movement on the level of motorneurons, Prof. dr. J. Voogd (Erasmus University Rotterdam) and Prof. dr. D. Kernell (University of Amsterdam) presented their views on these issues, respectively. Prof. dr. S. Grillner (Karolinska Institute Stockholm) resumed the afternoon session introducing the audience to neural networks and neural mechanisms in the lamprey. The symposium was concluded by Prof. dr. F.H. Lopes da Silva (University of Amsterdam) discussing new methods in research on the functioning of the brain.

      The structure of the book closely relates to the symposium program. The contributions presented in the morning form Chapters 1 and 2. They are followed by reactions of two panel members, one in collaboration with the chairman (Prof. dr. P.A. Huijing and Prof. dr. B. Hopkins & Prof. dr. A. Gramsbergen, all of the Vrije Universiteit Amsterdam). The same chapter organisation is applied to the contributions presented at the afternoon session (Chapters 4 and 5) which are followed by reactions of dr. E. Otten (University of Groningen), Prof. dr. C.C.A.M. Gielen (University of Nijmegen) and dr. O.G. Meijer, drs. A.P. Post, & drs. R. Bongaardt (Vrije Universiteit Amsterdam).

We would like to use this opportunity to gratefully acknowledge the help and support of a number of people. Albert Gramsbergen helped us to get in touch with the speakers and directed the day as chairman. Prof.dr F.H. Lopes da Silva introduced us to Prof. dr. Grillner. Olga Schipper en Arthur van der Meer were always willing to give practical and financial advice. The announcements were designed by Lia Out and at the symposium day we were assisted in several practical ways by Jacqueline van Adrichem, Fred Housheer, Raoul Oudejans, and Joeri van Wegen. The 'Stichting Het Vrije Universiteitsfonds' financially supported the symposium. Of course we would not have been able to organise the symposium and to edit this book without the contributions of both speakers and panel members, or without the financial freedom provided by the board of the Faculty of Human Movement Sciences.

The 1992 symposium organising committee,
Frank Zaal, Mark Willems, Lieke Peper,
Caroline Doorenbosch, Monique Berger, and
Christa Bakker.

Amsterdam, February 1993.

Permissions

We express our gratitude towards the Annual Reviews Inc for the permission to reproduce Grillner (1991) from the *Annual Review of Neuroscience*, **14**, 169-199.

In addition we thank the following publishers for allowing us to reproduce previously published figures:

Cambridge University Press. Figure 2, Chapter 2 (Eerbeek et al., copyright 1984).

Elsevier Science Publishing Company Ltd. Figure 1, Chapter 6 (Kandel & Schwartz, copyright 1985); Figure 5, Chapter 10 (Varela, copyright 1979).

MacMillan Publishers Ltd. Figure 7, Chapter 10 (Meyrand et al., copyright 1991).

Springer Verlag GmbH & Co. Figure 3, Chapter 6 (Georgopoulos, copyright 1983); Figure 4, Chapter 6 (Lang et al., copyright 1991); Figure 2, Chapter 10 (Cohen & Wallén, copyright 1980).

Universitá degli Studi di Pisa. Figure 1, Chapter 2 (Kernell, copyright 1992).

John Wiley & Sons Inc. Figure 2, Chapter 6 (Arbib, copyright 1989); Figure 6, Chapter 10 (Getting, copyright 1988).

Contents

Chapter 1: Anatomy of Motor Systems

J. Voogd

Motor systems can be arbitrarily subdivided into the following hierarchical levels:
1. Segmental motor systems, including motor- and interneurons and their peripheral input.
2. Subcortical motor centres located in the brain stem and their efferent motor pathways. They include Kuypers' medial and lateral motor systems with their topical projections to the spinal cord, and G. Holstege's 'third motor system', comprising the raphe nuclei and the locus coeruleus with their diffuse projections upon the cord.
3. Cortical motor centres: motor and premotor cortex and their cortico-spinal projections via the pyramidal tract.
4. Cerebellum and striatum: motor centres which act on the segmental motor system via subcortical motor centres and the motor cortex.
Anatomical and functional aspects of these levels in motor control will be discussed.

1 Introduction

Motor systems supply the input for the motorneurons, which innervate the skeletal muscles, and to visceromotor neurons, which connect with the neurons of the peripheral autonomous nervous system. Anatomy provides the nomenclature and the schemes of reference for the motor system. It may tell you where certain events take place and how they could be interrelated, but it cannot tell you how the central nervous system (CNS) works. In this contribution the different hierarchical levels of the motor system will be defined and the control will be discussed. A full account of the anatomy of the motor system can be found in Nieuwenhuys et al. (1988).

2 Levels in motor control

The motor system is characterised by a hierarchical organisation and by the presence of many parallel pathways connecting the different levels. I will arbitrarily distinguish four levels in the organisation of the motor system (Figure 1).

The first, or segmental level comprises the motoneurons, the interneurons (premotor neurons perhaps would be a better term) and their peripheral input. The wiring of the interneurons determines stereotyped motor behaviour, as in reflexes or rhythmic movements. Motoneurons and visceromotor neurons are found in the spinal cord and the brain stem, and the connections of their premotor neurons, therefore, involve large parts of the anatomy of these regions.

The second level includes the subcortical motor centres in the brain stem. They project to the motoneurons and the interneurons through long, ascending and descending pathways. One of these pathways is the medial longitudinal fascicle, which runs all the way along the brain stem and the spinal cord. Subcortical motor centres usually subserve some kind of motor strategy which involves different motor- and interneurons located at different levels of the CNS. Examples are the vestibular nuclei, which use information from the vestibular labyrinth to compensate for changes of the position of the head in space, this compensation occurs at the level of the spinal cord for postural adjustments and at the level of the brain stem for the compensation of the position of head and eyes; the superior colliculus (the 'tectum' with its descending motor system, the 'tectospinal tract') which coordinates gaze and associated head movements; centres for respiration and phonation which involve motoneurons innervating

Author's address: Department of Anatomy, Erasmus University Rotterdam, P.O.Box 1738, 3000 DR Rotterdam, The Netherlands.

many different muscle groups of the abdomen, pelvis, thorax, head and neck; the micturition centres which coordinate the innervation of the smooth muscle of the detrusor of the bladder and of the striated urethral sphincter and many others.

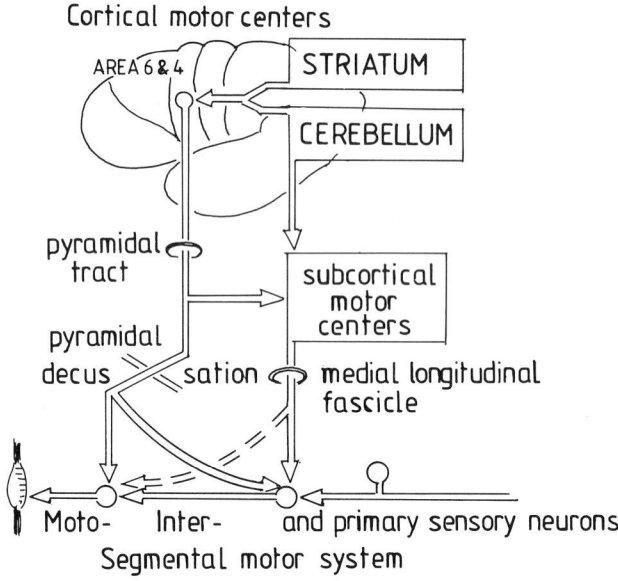

Figure 1. Diagram of the motor system. Subcortical and cortical motor centres are connected with the inter-and motoneurons of the segmental level through the medial longitudinal fascicle and the pyramidal tract respectively. Cerebellum and striatum exert their influence indirectly, through connections with the motor cortex and the subcortical motor centres.

The third level is the level of the motor cortex. The motor cortex is located in the frontal lobes of both hemispheres, just in front of the central sulcus. It contains an orderly arrangement of movements of different body parts and is used as a kind of keyboard in voluntary movement. Fibres from the motor cortex descend in the pyramidal tract. Most of these fibres cross and terminate on premotor and motor neurons in the brain stem and the spinal cord.

The fourth level includes the cerebellum and the striatum. The cerebellum is located in the roof of the fourth ventricle, the striatum is located in the lateral wall of the lateral ventricle and occupies the central part of the cerebral hemisphere. We know that lesions of these structures cause severe disturbances in motor behaviour, but we do not understand how or why this happens. Unlike the subcortical or cortical motor centres, the cerebellum and the striatum lack direct connections with the segmental level and unlike lesions of cortical and subcortical motor pathways, lesions of the striatum and the cerebellum do not give rise to a paralysis of movement. The striatum and the cerebellum maintain connections with certain subcortical motor centres and with the motor cortex, which mediate their effects on motor- and premotor neurons. The connections or diseases of the striatum and the cerebellum sometimes are indicated as 'extrapyramidal' connections or diseases. This obsolete term refers to a hypothetical pathway connecting the striatum with the motoneurons, which would run in parallel with the pyramidal tract. It will be clear that one of the major efferent routes of the striatum passes through the motor cortex and the pyramidal tract and does not circumvent it.

2

Parallel processing in the motor system means that a certain level may exert its influence on the segmental level through different, parallel routes, i.e., the motor cortex projects to motoneurons and interneurons through a direct, crossed pyramidal pathway but indirectly through several subcortical motor centres and through the cortico-cerebellar and cortico-striatal loops.

3 The segmental motor system

The simplicity of the wiring diagram of the segmental motor system in Figure 1 is misleading, the reality of the connections of the premotor neurons and their peripheral input is more complicated. One of the most simple sets of connections of spinal motoneurons is the wiring of the myotatic or stretch reflex (Figure 2). Large calibre, primary afferent fibres (Ia fibres), which enter the cord through the dorsal roots, subserve the afferent innervation of the muscle spindles, the receptors which register stretch in skeletal muscles. These fibres terminate with excitatory terminals[1] on the motoneurons innervating the same muscle. The net effect of the stretch reflex, therefore, is to keep the length of a muscle constant, irrespective of its load. Monosynaptic excitation of the agonist motoneurons is accompanied by inhibition of the antagonists. Inhibition is accomplished through an intercalated inhibitory interneuron (Ia inhibitory interneuron) which projects to the motoneurons innervating the antagonist muscles. The monosynaptic Ia innervation of the motoneurons can be suppressed by presynaptic inhibition[2] of these terminals by another set of inhibitory interneurons (Figure 3). Premotor networks, therefore, contain excitatory and inhibitory elements, which may use (combinations) of different neurotransmitters and which may establish axo-dendritic or somatic and axo-axonal connections.

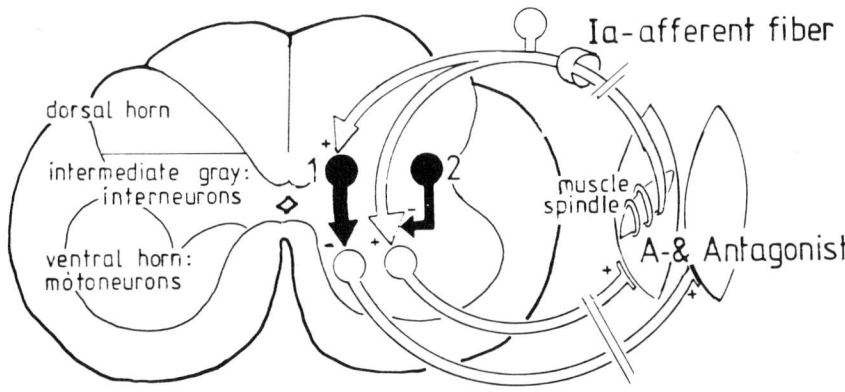

Figure 2. Motor- and interneurons in the spinal cord. Motoneurons are located in the ventral horn; interneurons occupy the intermediate grey. The wiring of the stretch reflex is illustrated: Ia afferent fibres which subserve the afferent innervation of the muscle spindle, terminate with excitatory terminals on motoneurons innervating the same muscle, and on an Ia inhibitory interneuron (1) which inhibits the motoneurons innervating the antagonist. Another type of inhibitory interneuron (2) terminates with presynaptic boutons on the Ia terminals on motoneurons, and thus blocks Ia-motoneuronal transmission (pre-synaptic inhibition).

---

[1]. Excitatory and inhibitory terminals, i.e. terminals containing excitatory or inhibitory neurotransmitters, can be distinguished under the electron microscope under certain standard conditions, by their contents of spherical and flattened synaptic vesicles respectively.

[2] Presynaptic inhibition involves an axo-axonal synapse. In this case GABAergic synapses are present on the excitatory terminals of the Ia fibers on the motoneurons.

3

4 Interconnections between levels of the motor system: The medial and lateral subcortical motor systems of Kuypers

Important contributions to our understanding of the subcortical and cortical motor systems were made by the late Professor Kuypers, professor of anatomy in Rotterdam and, since 1985, in Cambridge, and his pupils, Dr.G. and Dr.J.C. Holstege, at present in the Departments of anatomy in Groningen and Rotterdam respectively. Kuypers distinguished between medial and lateral subcortical motor pathways, which descend in the ventral and lateral funiculus of the cord and terminate on medially and more laterally located motor- and interneurons respectively. Medially located inter- and motoneurons innervate axial and proximal muscles; laterally located inter-and motoneurons subserve more distal movements of the limbs. Motoneurons innervating the flexors and the distal hand- and foot muscles are located in the extreme dorsolateral part of the ventral horn (Figure 4). Lateral subcortical motor pathways, such as the rubrospinal tract, which takes its origin from a group of neurons (red nucleus) in the mesencephalon, and which crosses at this level, influence distal movements of the limbs (Figure 5A). Medial pathways, such as the vestibulospinal and tectospinal projections through the medial longitudinal fascicle (Figure 5B), exert a bilateral influence on proximal and axial muscles (i.e. the muscles involved in postural stability and in gaze-associated movements of the head).

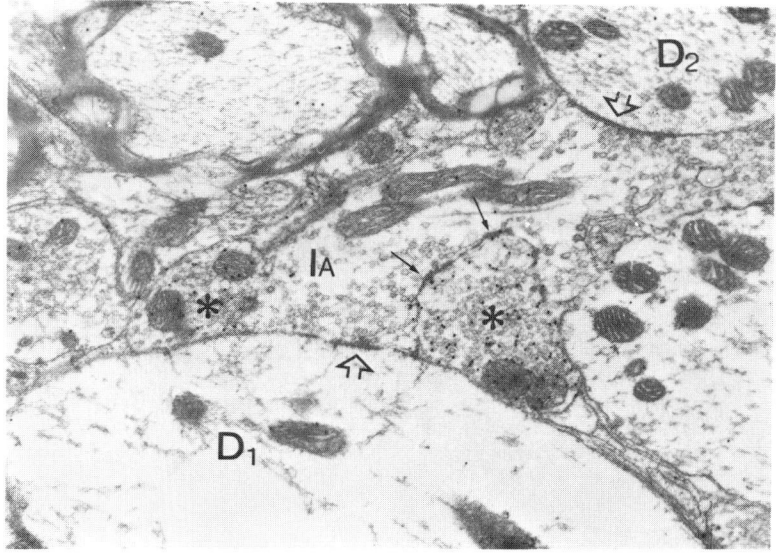

Figure 3. EM micrograph of an Ia terminal (Ia) synapsing on two dendrites of motoneurons (D1, D2, open arrows). An inhibitory terminal (asterisk: the grains on this terminal indicate the presence of the inhibitory neurotransmitter gamma aminobutyric acid -GABA-) synapses with the Ia terminal (small arrows: presynaptic axo-axonal terminal). Courtesy of Dr. J.C. Holstege, Rotterdam.

5 The pyramidal tract

Kuypers showed that the pyramidal tract terminates on both populations of medial and lateral inter- and motoneurons (Figure 5A and B). However, the cortico-motoneuronal connections to laterally located motoneurons which subserve fine finger and hand movements take their origin from 'hot spots' on the motor cortex, representing distal movements and characterised by the presence in the cortex of collections of large neurons (giant pyramidal cells of Betz). Cortico-motoneuronal projections to proximal and axial muscles are bilateral, and arise from

4

more peripheral, and more diffusely organised regions of the motor cortex. The effects of lesions of the pyramidal tract on motor performance differ for different species and with their age. Interruption of the pyramidal tract in ungulates, where the pyramidal tract does not descend beyond the high cervical cord, and in carnivores and rodents, where the crossed pyramidal tract descends in the lateral or the dorsal funiculus till sacral levels, does not lead to motor disturbances. In monkeys lesions of the pyramidal tract cause a permanent loss of the ability to perform fine distal movements (precision grip) of the extremities. In humans it leads to a crossed spastic hemiplegia. When the lesion is present at birth the paralysis and the spasticity develop some time later, presumably at the time when the connections of the pyramidal tract should have become established.

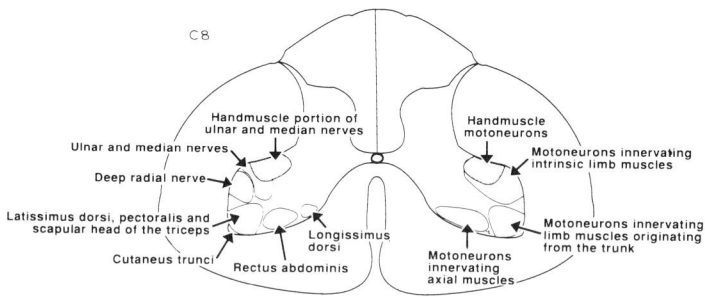

Figure 4. Diagram of the motoneuronal cell groups in the ventral horn at C8 in the cat. From G. Holstege (1991).

The lack of motor symptoms in lower mammals and in neonatal primates with lesions of the pyramidal tract and the development of spasticity in humans but not in other species under these conditions, remain difficult to explain. The relative lack of direct cortico-motoneuronal connections is often used to explain the lack of motor symptoms. The presence of direct or indirect excitatory connections of the pyramidal tract with the inhibitory interneurons which suppress the Ia excitatory input to the motoneurons has been assumed to explain the spasticity (Figure 2 and 3).

6 Limbic influences on motor systems

G. Holstege (1991) coined the term 'third motor system' for the ensemble of the limbic system and the hypothalamus which project, through the central grey, the raphe nuclei and the locus coeruleus to the segmental motor system (Figure 6). The projections of the nucleus raphe magnus and the adjacent reticular formation and of the locus coeruleus to the segmental motor system are diffuse. They lack the focusing of Kuypers' lateral subcortical motor system and of the subdivisions of the corticospinal tract. According to Holstege, these systems do not necessarily induce movement, but they set the general level of excitability of the motor system. During conditions of stress the serotoninergic projections from the nucleus raphe pallidus, located at rostral levels of the medulla oblongata, would increase the excitability of the motor system, while more rostral subdivisions of the raphe nuclei (nucleus raphe magnus) would depress pain conduction through the dorsal horn and the spinal trigeminal nucleus. J.C. Holstege showed that the situation is more complicated and that long inhibitory descending pathways, containing glycine and GABA (which may be co-localised with serotonin) take their origin from the ventromedial reticular formation adjacent to the nucleus raphe pallidus and raphe magnus and terminate on motoneurons (Holstege, 1987, 1991; Holstege and Calkoen, 1990; Holstege and Bongers, 1991).

5

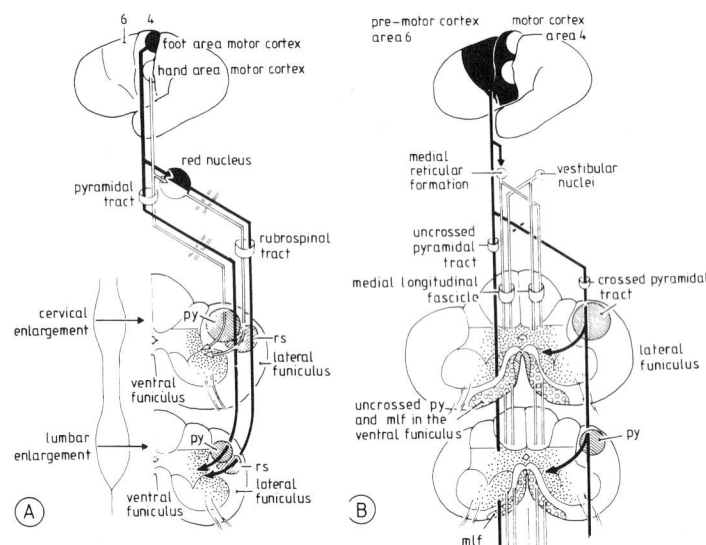

Figure 5A. Lateral subcortical motor system of Kuypers takes its origin from the red nucleus, crosses the midline and descends as the rubro-spinal tract (rs) in the lateral funiculus of the cord. It terminates on laterally located inter- and motoneurons. Fibre contingents of the pyramidal tract (py) arising from foot and hand areas of the motor cortex, cross and also descend in the lateral funiculus. Both the rubrospinal tract and the corticospinal pathway are somatotopically organised and both terminate in the same, lateral regions of the spinal cord.

Figure 5B. Medial subcortical motor systems take their origin from the medial reticular formation and the vestibular nuclei. They descend bilaterally in the medial longitudinal fascicle (mlf) in the ventral funiculus of the cord to terminate bilaterally on medially located inter- and motoneurons. Corticospinal projections from the rostral motor cortex descend in the pyramidal tract and partially decussate. Uncrossed fibres descend in the ventral funiculus, crossed fibres in the lateral funiculus. These corticospinal fibres terminate bilaterally in the same, medial regions of the spinal grey matter.

7 Striatum and cerebellum

The anatomical relations of the striatum and the cerebellum with the motor cortex are rather similar (Figure 7). Both receive a projection from the entire cerebral cortex, including the motor cortex. The striatum receives direct cortico-striatal fibres, the connections to the cerebellum pass through the nuclei pontis, which are located in a bulge - the pons -surrounding the pyramidal tract at the ventral aspect of the brain stem. Both the striatum and the cerebellum give rise to efferent projections which converge upon the frontal lobe of the hemisphere, which contains the motor cortex. The efferent path of the striatum includes the globus pallidus and the ventral thalamus. The striato-pallidal and the pallido-thalamic links are both inhibitory. The net effect of excitation of the striatum, therefore, is disinhibition (i.e., an increase in excitability) of the thalamus and the frontal cortex. The efferent cerebello-thalamic-cortical pathway is excitatory. The cerebellum converges upon the motor cortex, the cortical region immediately in front of the central sulcus; the projections of the stratum are located more anteriorly, in the so-called premotor and prefrontal cortex. The latter regions are involved with initiative and the preparation of movement.

The function of the striatum remains difficult to imagine. Parkinson's disease, which affects the striatum through the degeneration of the substantia nigra, a nucleus which provides the striatum with dopamine, is characterised by involuntary movements in the form of the

Parkinsonian tremor which occurs during wakefulness, hypokinesia and rigidity. Patients have difficulty in starting or changing movements, but sometimes are able to execute apparently normal movements under visual control - catching a ball - or when under stress. Dopamine supposedly suppresses the activity of striatal neurons and thus causes a disinhibition of the globus pallidus, leading to a lower excitability of the premotor and prefrontal cortex (Delong, 1990). The hypokinesia of Parkinson's disease, therefore, may be related to the function of these cortical fields. Other efferent pathways of the striatum are directed at the tectum and on a region in the mesencephalon, known as the mesencephalic locomotion centre. They are involved in the control of eye movement and in the initiation of walking. Dopamine-induced changes in the capacity of striatal neurons to synthesise different neurotransmitters, are an important subject of research in the Department of Anatomy of the Free University in Amsterdam.

Anatomically the cerebellum can be considered as a side-loop of many motor systems. The cerebellum is endowed with a cortex, which is build according to strict geometrical principles (Figure 8A). The output cells of the cerebellar cortex, the Purkinje cells, are arranged in a number of parallel, parasagittal strips (i.e., bands which run from the anterior to the posterior margin of the cerebellum). Each of these strips monitors the activity of a certain motor centre (for instance one of the medial or lateral subcortical pathways of Kuypers or the motor cortex). Each strip is provided by a set of afferent 'climbing' fibres from the inferior olive, a nucleus in the medulla oblongata. These fibres derive their name from the fact that they climb into the para-sagitally oriented dendritic trees of the Purkinje cell. The main afferent

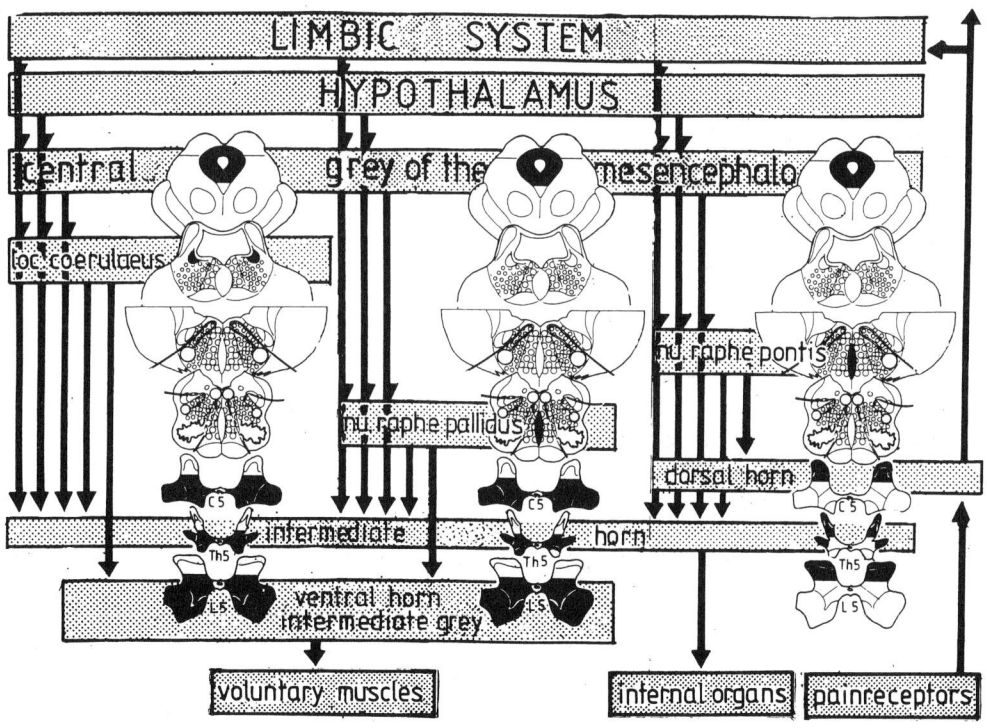

Figure 6. Diagram illustrating G. Holstege's '3rd motor system' with its diffuse connections of the raphe nuclei and the nucleus of the locus coeruleus with the spinal cord. Supraspinal centres which influence the raphe nuclei and the nucleus of the locus coeruleus include the limbic system, the hypothalamus and the central grey.

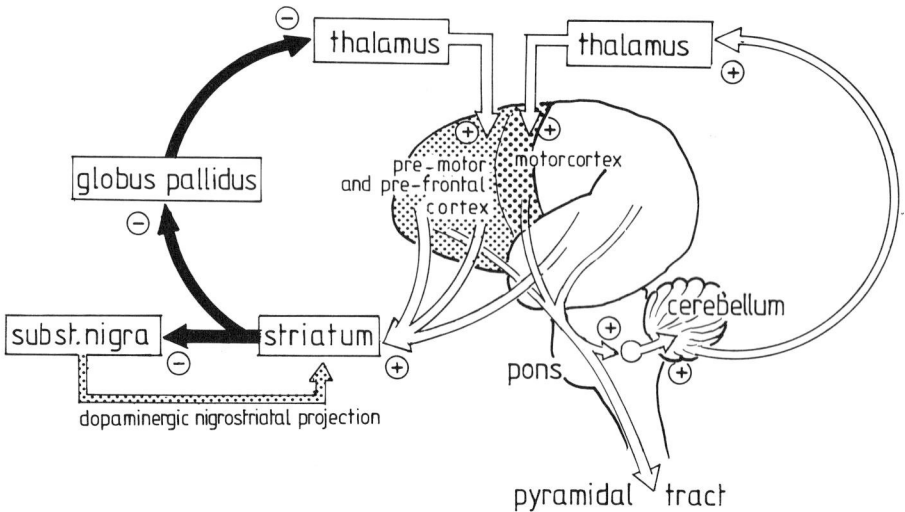

Figure 7. Diagram of the cortico-striato-cortical and cortico-cerebello-cortical 'loops'.

systems (the mossy fibre - parallel fibre or MP pathways) to the cerebellar cortex are oriented transversely, perpendicular to the Purkinje cell strips. This arrangement favours the chance that all afferent MP system, whatever their source, have access to all the output-strips of the cortex and, consequently, to the motor systems which they monitor. One of these MP systems, discussed in the previous paragraph, is the cortico-ponto-cerebellar pathway, which provides the cerebellum with a copy of the activity in the pyramidal tract. Diseases involving the cerebellum cause a lack of coordination of movement, known as cerebellar ataxia, which resembles the motor disturbances following alcohol intoxication. The ataxic walk, mistakes in the calibration of movements of the limbs and the eyes, dysarthria and hypotonia are typical of both conditions.

Rigid explanations of the cerebellar syndrome are not available, but many theories abound. Some of these theories have in common that they attribute to the cerebellum a role in motor learning. They postulate that simultaneous activity in the climbing fibre and the MP pathway will cause a permanent change in the effectivity of the MP - Purkinje cell synapses (Marr, 1969; Albus, 1971). Because the cerebellum forms a side loop of the 'main' motor pathway (i.e. the cortico-ponto-cerebellar path which detaches from the main, pyramidal system in the pons, Figure 7) the ultimate effect of the main and the cerebellar path to the moto-neurons will be the sum of the activity in both pathways (i.e. the primary effect of the pyramidal tract fibres on motoneurons plus the indirect effects of the cerebellum, through the cortical and subcortical motor centres). Changes in the effectivity of the MP-Purkinje cell synapse will enhance or diminish the contribution of the cerebellum to this sum. One theory states that climbing fibres carry an error signal. Each time a mistake occurs in the calibration of movements, these errors are signalled by climbing fibres and lead to plastic changes in the MP-Purkinje cell synapses till the calibration of the movement is restored to its desired value (Figure 8B). The cerebellum learns from its mistakes.

Another theory states that the cerebellum is essential in the acquisition of certain conditioned reflexes (Yeo et al., 1985A, B and C). Climbing fibres activate certain parasagittal Purkinje cell strips during the unconditional stimulus (Figure 8C). Conditional stimuli activate different MP pathways. Repeated combinations of the conditional and the unconditional stimulus causes a permanent change in the MP-Purkinje cell synapses located at the cross

8

roads of the transversely oriented MP pathway and the parasagittally dis-positioned Purkinje cell strip. In due time the conditional stimulus may activate the appropriate motor pathway through its MP pathway and the changed Purkinje cells. In the first 'learning theory' of the cerebellum the climbing fibre, carrying the error signal, is the main variable. In the second theory the main variable is the MP system carrying the conditional stimulus. The learning theories of the cerebellum have not remained uncontested, but received some strong experimental support.

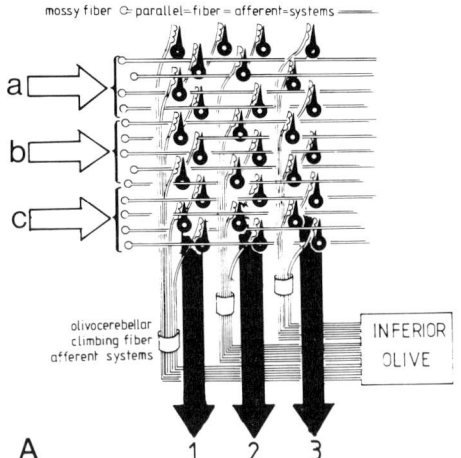

Figure 8A. Diagram explaining the 'lattice' structure of the cerebellar cortex. Purkinje cell dendrites are arranged in a parasagittal plane; Purkinje cells themselves are organised in parasagittal strips. The output from each Purkinje cell strip monitors a certain motor system (1,2 or 3). The input of the cortex consists of two systems: transversely mossy fibre-parallel systems (a, b and c) terminate on Purkinje cell dendrites. Each Purkinje cell strip is provided with its own climbing fibre afferent system.

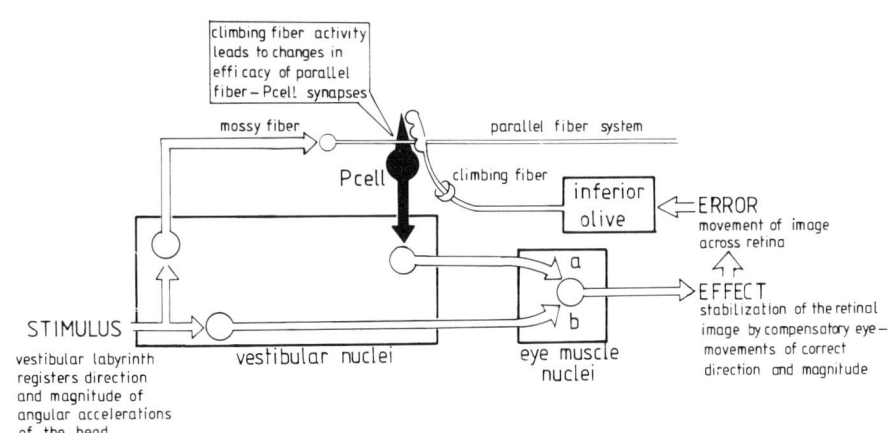

Figure 8B. Diagram illustrating motor learning by the cerebellar cortex. Reflex (in this case the vestibulo-ocular reflex) connections induce a certain, stereotyped effect (in this case a compensatory movement of the eye which stabilises the retinal image). The cerebellar cortex is part of a superimposed loop. The effect is the sum of the signal transmitted by the reflex path (b) and the cerebellar loop (a). Signal a can be adjusted, because, according to the theories of Marr (1969) and Albus (1971), simultaneous activity in climbing fibres and parallel fibres, modifies the efficacy of the parallel fibre-Purkinje cell transmission. In the case of the vestibulo-ocular reflex, the olivocerebellar climbing fibre pathway carries an error signal (Ito, 1984, i.e. the image is not stabilised, but moves across the retina). Changes in the parallel fibre-Purkinje cell synapse, induced by the climbing fibre, modify signal a and rectify the effect.

9

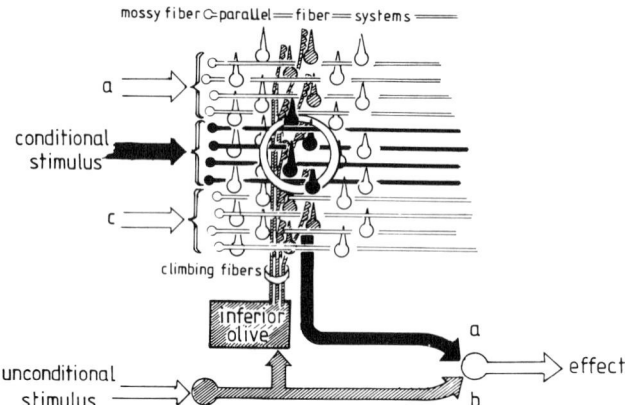

Figure 8C. Diagram illustrating another type of motor learning, i.e., the establishment of conditioned reflexes by the cerebellum. The unconditioned stimulus gives rise to a certain effect through pathway *b*. A side path to the inferior olive stimulates climbing fibres terminating on a parasagittal strip of Purkinje cells. Any conditioned stimulus leads to the activation of an afferent mossy fibre-parallel pathway. Simultaneous activity in the unconditioned (climbing fibre) and conditional (mossy fibre-parallel fibre) pathways results in permanent changes in the efficacy of the parallel fibre-Purkinje cell synapses on Purkinje cells located at the intersection of both pathways (circle). After 'learning' has been completed the conditional stimulus leads to the same effect through the 'changed' Purkinje cells and pathway *a*.

Coordination of movement by the cerebellum, therefore, takes some time to become established. Once it is there, it works perfectly, till the circuits are disturbed by alcohol or disease.

Motor systems are complicated. Advances in our understanding may result from careful observations of movements and movement disorders, by systematic analysis of the anatomy, physiology and pharmacology of motor circuits in experimental animals, by the design of experimental techniques to test the results of these experiments in humans, but foremost by the development of new, comprehensive theories on the steering of movement by the brain.

References

Albus, K. (1971). A theory of cerebellar function. *Bioscience*, **10**, 25-61.

Delong, M.R. (1990). Primate models of movement disorders of basal ganglia origin. Trends in *Neuroscience*, **13**, 281-285.

Holstege, J.C. (1987). Brainstem projections to lumbar motoneurons in rat II. An ultrastructural study by means of the anterograde transport of wheat-germ agglutinin coupled to horseradish peroxidase and using the tetramethyl benzidine reaction. *Neuroscience*, **21**, 368-376.

Holstege, J.C. (1991). Ulstrastructural evidence for GABAergic brain stem projections to spinal motoneurons in the rat. *Journal of Neuroscience*, **11**, 159-167.

Holstege, J.C., & Calkoen, F. (1990). The distribution of GABA in lumbar motoneuronal cell groups. A quantitative ultrastructural study in rat. *Brain Research*, **530**, 130-137.

Holstege, J.C., & Bongers, C.M.H. (1991). A glycinergic projection from the ventromedial lower brain stem to spinal motoneurons. An ultra-structural double labeling study in rat. *Brain Research*, **566**, 308-315.

Holstege, G. (1991). Descending motor pathways and the spinal motor system: Limbic and non-limbic components. In G. Holstege (Ed.), *Role of the forebrain in sensation and behavior: Vol. 87. Progress in Brain Research:* (pp. 307-421).

Ito, M. (Ed.) (1984). The cerebellum and neural control. New York: Raven Press.

Kuypers, H.G.J.M. (1981). Anatomy of the descending pathways. In J.M. Brookhart, V.B. Mountcastle, V.B. Brooks & S.L. Geiger (Eds.), *Handbook of Physiology, The Nervous System: Vol. II: Motor Control part1* (pp. 597-666). Bethesda MD: American Physiological Society.

Marr, D. (1969). A theory of cerebellar cortex. *Journal of Physiology*, **202**, 437-470.

Nieuwenhuys, R., Voogd, J., & Van Huijzen, Chr. (1988). *The Human Central Nervous System. An Synopsis and Atlas* (3rd rev. ed.). Berlin-Heidelberg-New York-London-Paris-Tokyo: Springer-Verlag.

Yeo, C.H., Hardiman, M.J., & Glickstein, M. (1985a). Classical conditioning of the nictitating membrane of the rabbit. I. Lesions of the cerebellar nuclei. *Experimental Brain Research*, **60**, 87-98.

Yeo, C.H., Hardiman, M.J., & Glickstein, M. (1985b). Classical conditioning of the nictitating membrane of the rabbit. II. Lesions of the cerebellar cortex. *Experimental Brain Research*, **60**, 99-113.

Yeo, C.H., Hardiman, M.J., & Glickstein, M. (1985c). Classical conditioning of the nictitating membrane response of the rabbit. III. Connections of cerebellar lobule HVI. *Experimental Brain Research*, **60**, 114-126.

Chapter 2: The gradation of muscle force and the functional organisation of spinal motoneurons and motor units

D. Kernell

A precise gradation of muscle force is of utmost importance for motor control in general. In the present contribution a brief survey has been given of the mechanisms involved at the level of the "output interface" of the central nervous system: the alpha motoneurons and their muscle units. The questions dealt with included:

1) mechanisms of relevance for the property-ranked recruitment hierarchy of motoneurons, i.e. recruitment gradation with the neurons becoming activated, statistically speaking, in a sequence related to the contractile properties of their muscle fibres (cf. Henneman's 'size principle');

2) the possible relation between the topography of motoneurons and muscle units and the variations in precise recruitment pattern that tend to take place between different motor tasks (task-related recruitment strategies);

3) mechanisms and biophysical motoneuron properties of relevance for the rate-gradation of force;

4) the importance for force gradation of an appropriate matching between relevant motoneuronal and muscle unit properties (concerns both the recruitment and rate gradation of force);

5) the extent to which such matches might be maintained (and, if needed, adjusted) by long-term effects of motoneuronal activity patterns on muscle fibre contractile properties.

1 Introduction

One of the main requirements for a well-coordinated motor behaviour is that the central nervous system (CNS) must be able to produce (and, if necessary, maintain) the correct contractile force in all the muscles concerned. The forces produced may be needed for joint stabilisation (posture), movement production (shortening contraction) or movement braking (lengthening contraction). As each mammalian muscle is controlled by a *population* of motoneurons (i.e. by a multi-unit 'interface'), the CNS can grade force by two alternative mechanisms which are often used in parallel: 1) by varying the number of activated motor units (*recruitment gradation*), and 2) by varying the discharge rate of already recruited units (*rate gradation*). In the present chapter, I will give a brief but fairly general survey of current problems and observations concerning both gradation mechanisms. In this context, I will also deal with the long-term effects of different motoneuronal activity patterns on contractile muscle properties, a problem of relevance for long-term aspects of neuromuscular force gradation. My account will be centred around, and illustrated by, findings from our own work on hind limb motoneurons and motor units from experimental animals (mainly cat). The main principles are, however, as far as we know applicable to the human motor system as well.

2 Recruitment gradation

2.1 Recruitment strategies

In the majority of voluntary or reflex contractions that have been studied, the most easily activated alpha motoneurons tend to be those with the thinnest axons. In accordance with the *size-principle of Henneman* (Henneman & Mendell, 1981), a progressive increase of force is, statistically speaking, produced by the activation of progressively more thick-axoned cells. One of the reasons for the functional importance of this 'ascending-size-order of recruitment'

Author's address: Department of Neurophysiology, University of Amsterdam, Academisch Medisch Centrum, Meibergdreef 15, 1105 AZ Amsterdam, The Netherlands.

lies in the fact that axonal size commonly correlates with essential contractile properties of the motor units (Burke, 1981; Henneman & Mendell, 1981): the thinnest axons tend to innervate units that are relatively slow, weak and fatigue-resistant. Conversely, the thickest motor axons, which are more difficult to recruit, tend to have units that are fast and strong. Thus, the ascending-size-order of recruitment represents a case of what one might also more generally term a *property-ranked recruitment hierarchy*. Even though some properties of motoneurons and muscle units vary independently of axonal conduction velocity (e.g., differences between fast-twitch units with regard to their fatigue-resistance, motoneuronal input resistance and motoneuronal excitability: Burke et al., 1973; Fleshman et al., 1981; Kernell & Monster, 1981), axonal and other aspects of neuronal size are without any doubt essential factors in the quantitative and statistical analysis of the organisation and behaviour of motoneuron - muscle unit populations. The size principle may be seen as a valuable summarising concept (a 'conceptual tool'), pointing to the important fact that so many of the properties of motoneurons and muscle units are mutually interrelated and co-varying with aspects of neuronal size (cf. Henneman & Mendell, 1981). There remain, however, many intriguing questions (Enoka & Stuart, 1984), including the problem of what the possible *causal* links are between motoneuronal size and the various functional properties implicated in the size principle (cf. Lüscher et al., 1979).

There is much evidence indicating that, besides the property-ranked recruitment hierarchy there is also a *task-related recruitment strategy*. Thus, even among units that exert their force in the same direction, different motor tasks often seem to show a preference for different individual units within the same muscle (e.g. Desmedt & Godeaux, 1981; ter Haar Romeny et al., 1984; Hoffer et al., 1987). The two main types of recruitment strategy that I have mentioned should *not* be seen as mutually exclusive alternatives, but most probably they are both active in parallel. According to the property-ranked strategy the small-axoned and slow units would be those most easily activated in the majority of motor tasks. The task-related strategy might then determine precisely which ones of the various slow and small-axoned units would be those preferentially activated in a given task.

After these introductory comments concerning recruitment strategies, I would like to proceed to discuss some of the mechanisms involved.

2.2 Motoneuronal membrane properties and size-related recruitment strategies

The ease with which an individual motoneuron within a pool may become activated during a given motor act must depend on the combination of at least two groups of factors:
1) the extent to which the cell is favoured by that particular set of active synapses, and
2) the ease with which the cell is fired by excitatory synaptic current.
As the ascending-size-order of recruitment occurs so often and in combination with so many different types of synaptic organisation (Henneman & Mendell, 1981), the second factor becomes particularly interesting: do small-axoned slow motoneurons have membrane properties that make them more easily excited than larger and faster cells?

Already a long time ago, it was found that motoneurons with a slow axonal conduction velocity also had a high input resistance and a low current threshold for the activation of repetitive impulse firing (Kernell, 1966). The differences in current threshold were probably to an important extent caused by the differences in input resistance: if the various cells had to become depolarised to about the same extent to start firing, then less current would obviously be needed for discharging those with a high input resistance than for those with a lower one (cf. Ohm's law). Originally, these differences were thought to have a fairly simple explanation (Kernell, 1966): they could have been produced by differences in soma-dendrite size. As was demonstrated by Cullheim (1978) as well as in our own laboratory (Kernell & Zwaagstra, 1981), motoneurons with thin axons do indeed also tend to have a relatively small soma. Furthermore, motoneurons with a small soma tend, on average, to possess relatively thin proximal dendrites (Kernell, 1966; Zwaagstra & Kernell, 1981; Ulfhake & Kellerth, 1981). If other conditions remain constant, a small membrane area would present a greater electrical resistance than that of a larger area. Continued experimental analysis demonstrated, however, that size was not the only factor involved. Combined anatomical and physiological measurements on the same motoneurons showed that the specific membrane resistance (membrane resistivity) was higher for the small-axoned slow motoneurons than for faster cells with larger axons (Kernell & Zwaagstra, 1981, 1989; Ulfhake & Kellerth, 1984; Fleshman et al., 1988; see also Gustafsson & Pinter, 1985).

The existence of systematic differences in membrane resistivity between large- and small-axoned motoneurons has important consequences for the potential reflex excitability of these various cells. Even if a given class of activated excitatory synapses were randomly distributed among the cells of a motoneuron pool, the differences in specific membrane resistivity would cause a greater depolarisation to be produced in small-axoned cells than in those with larger axons (Kernell & Zwaagstra, 1981). Hence, even in the absence of any specificity in the distribution of synapses within a motoneuron pool, the ascending-size-order of recruitment would tend to appear. We feel that these results help to explain why the ascending-size-order of recruitment is so commonly encountered in motor physiology.

Differences in membrane resistivity are also likely to be largely responsible for the fact that, for the cat's triceps surae, motoneurons of fast-twitch fatigue-resistant units (FR) are electrically more excitable than those of the fast-twitch fatigue-sensitive ones (FF; Fleshman et al., 1981; Kernell & Monster, 1981). These two classes of motoneurons do not differ significantly in axon or soma size (Burke et al., 1973; Ulfhake & Kellerth, 1982), but the input resistance is higher for FR than for FF units (Fleshman et al., 1981).

Besides being dependent on membrane resistivity, the electrical excitability would also be influenced by subthreshold rectification processes as well as by a variation in voltage threshold and/or in resting membrane potential. Although some studies indicate that there may indeed exist systematic differences in the amount of threshold depolarisation required for motoneurons of different functional type (Fleshman et al., 1981; Gustafsson & Pinter, 1984), such differences might commonly be less marked than those referring to membrane resistivity (cf. Pinter et al., 1983).

2.3 Motoneuronal soma position and property-ranked recruitment strategies

Our conclusion that differences in membrane properties are of importance for recruitment behaviour does not, of course, mean that this is the only factor involved. Differences in synaptic distribution as well as in synaptic activation-efficiency (cf. Lüscher et al., 1979) are also likely to be of great importance. There is ample evidence indicating that various sets of synapses exist that may tend to sharpen or neutralise, or even partly reverse, the apparently intrinsic tendency for an ascending-size-order of recruitment (e.g., Burke et al., 1970; Garnett & Stephens, 1981). Central synapses are often (but not at all always) organised on topographical principles. Hence, we have been interested in finding out to what an extent the spatial organisation of motoneurons and motoneuron pools might be of relevance for recruitment strategies. For instance, we wondered whether motoneurons that belonged to different categories with respect to their property- or task-related recruitment would tend to lie in different portions of the motoneuron pool. We have investigated such questions for the motoneurons innervating the peroneus longus muscle (PerL) of the cat's hind limb. This muscle is suitable for such studies concerning central aspects of motoneuronal organisation, because peripherally all the PerL muscle units are mechanically equivalent with respect to force-direction: they are all connected to the foot via the same long and slender tendon.

The results of these investigations showed that the length-wise distribution of motoneurons with different 'types' of muscle fibres was not completely random: cells with slow and fatigue-resistant muscle fibres were slightly but significantly more common in caudal

14

than in more cranial portions of the motoneuron pool (Kernell et al., 1985). The non-random distribution was, however, not distinct enough to serve as a topographical basis for a type-specific motoneuronal innervation: fast and slow units were intermingled in all parts of the pool. The non-randomness of this intermixture might, however, suggest that rostral and caudal PerL units were subjected to different types of daily use (see below).

2.4 Motoneuronal soma position and the site of the innervated muscle fibres

For many muscles, there is an evident correlation between the rostro-caudal position of a motoneuron within its spinal pool and the intramuscular site of its nerve endings (e.g., Swett et al., 1970). We have used the technique of glycogen depletion for investigating this relationship in the cat's PerL (Donselaar et al., 1985). We then found that rostral portions of the PerL motoneuron pool distributed their endings predominantly to anterior muscle portions, and vice versa. This meant that we could use localised electromyographic (EMG) recordings from anterior and posterior muscle portions for studying, in an indirect way, the distribution of activity between rostral and caudal portions of the intraspinal motoneuron pool. In recent experiments, we have used this approach for investigating whether different synaptic inputs to the motoneurons of a given muscle would tend to distribute their excitation to different rostro-caudal portions of the pool.

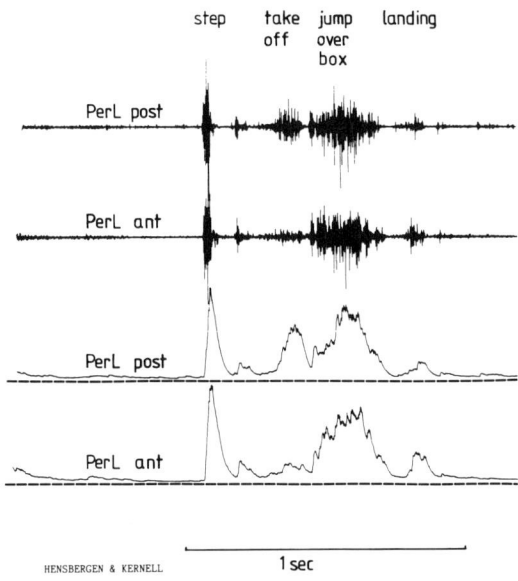

Figure 1. Differences in distribution of intramuscular discharge during different motor actions. Electromyographic (EMG) recordings from anterior and posterior regions of cat's PerL during voluntary unrestrained movements.
Upper two traces show raw EMG recordings and lower two traces show the same signals after rectification and smoothing ('integrated' EMG). The cat took a step and then jumped over a small box lying on the floor (height of box: 20 cm). Note the varying relationships between anterior and posterior EMG during the different phases of motor behaviour; posterior EMG dominates during jump take-off and anterior EMG during the landing from the jump. The EMG recordings were obtained from bipolar thin-wire electrodes with a relatively restricted recording area, and there was no significant cross-talk from adjoining muscles (controlled by simultaneous EMG recordings from the neighbours). From experiments of Hensbergen & Kernell (1992a); illustration reproduced from Kernell (1992).

15

2.5 Muscle unit site and task-related activation strategies

A first series of these experiments was performed in cats anaesthetised with pentobarbitone. In these animals, the antero-posterior distribution of PerL activity was significantly different depending on whether the muscle was activated via a flexion reflex or by electrical stimulation of the motor cortex (Kandou & Kernell, 1989). In subsequent explorations these experiments have been continued in freely moving cats and, again, systematic and significant differences in the intramuscular distribution of PerL activity was found depending on the motor task (Hensbergen & Kernell, 1992a). In, for instance, a cat jumping over a box lying on the floor, posterior PerL was relatively more dominant during take-off and anterior PerL more during landing (Figure 1). Considering the somatotopic relationships between PerL motoneurons and their muscle fibres (see above), such results strongly suggested that synaptic inputs associated with these different tasks may also differ systematically from each other with respect to their rostrocaudal distribution within the spinal cord. Thus, at least for the PerL muscle of the cat, topographical factors do indeed seem to be of importance for task-related aspects of motoneuronal activation (cf. also ter Haar Romeny et al., 1984).

3 Rate gradation

3.1 Tension-rate relation of muscle units

In skeletal muscle fibres, isometric force increases with activation rate according to a markedly sigmoid curve (Cooper & Eccles, 1930). The steep intermediate region of this *tension-frequency curve* is positioned at lower rates for motor units and muscles with a slow twitch than for those with a faster twitch (Cooper & Eccles, 1930; Kernell et al., 1983b). These isometric aspects of muscle 'speed' are probably mainly related to the kinetics of sarcoplasmic calcium movements (cf. Kugelberg & Thornell, 1983). How are the properties of motoneurons matched to their tasks in the rate modulation of muscle tension?

3.2 Frequency-current relation of motoneurons

As motoneurons become activated during a maintained contraction, they typically discharge at fairly regular intervals. Such maintained repetitive discharges are evoked by persisting activating currents, produced by the summation of many asynchronous postsynaptic events. The discharge-generating action of these postsynaptic currents may be well imitated by currents that are injected through the tip of an intracellular micro electrode (for review: Kernell, 1984). Experiments performed by aid of this technique have demonstrated a number of different ways in which the repetitive properties of motoneurons are well matched to the response properties of their muscle fibres. Main points are enlisted and briefly commented upon below:

1) *Minimum discharge rate.* In the discussion of rate gradation it is important to realise that motoneurons do not only have a maximum possible rate of discharge, but also a minimum one below which no regular firing is maintained (Kernell, 1965). As motoneurons are gradually recruited during a weak contraction of slowly increasing strength, each cell starts firing regularly at its own characteristic minimum rate, which tends to be similar for cells with a similar recruitment threshold. As contractile force is enhanced, firing rate also typically tends to increase for already discharging units (e.g., Kernell & Sjöholm, 1975). The minimum firing rate of a motoneuron is typically such that it corresponds to the lower end of the steep region of the tension-frequency relation of its motor unit (Kernell, 1965, 1984; Kernell & Sjöholm, 1975). Thanks to this *rate-match* between motoneurons and muscle fibres, an increase of discharge rate of an already recruited motoneuron will automatically lead to a significant increase in motor unit force.

16

The minimum rate of regular motoneuronal firing is commonly mainly determined by the time course of the post-spike after-hyperpolarization (AHP): at the minimum frequency, the impulse intervals are similar to the total duration of this afterpotential (Kernell, 1965). The rate-match between motoneuron and muscle is largely achieved by a match between the time courses of the motoneuronal AHP and the motor unit twitch: both phenomena are typically of about the same duration (Bakels & Kernell, in press).

2) *Maximum discharge rate.* The maximum possible impulse frequency that a motoneuron may maintain (e.g., during at least 0.5-1 sec) is not determined by its AHP but must depend on the spike-generating properties of its membrane (proneness for inactivation of voltage-dependent sodium-channels?). Earlier studies have shown that motoneurons may indeed commonly be made to discharge at rates high enough for producing a tetanic tension close to the maximum one of their muscle fibres (Kernell, 1965, 1984). Also this maximum rate is higher for AHP-fast than for AHP-slow motoneurons (maximum rate of 'secondary range'; Kernell, 1965, 1984).

3) *Shape of frequency-current curve.* At the very highest submaximal forces of a muscle, the tension-frequency relation becomes less steep. This decrease in tension-frequency slope is partly compensated for by the fact that, at such high rates of discharge, the frequency-current slope (f-I slope) of motoneurons tends to become steeper ('secondary range'; Kernell, 1984).

4) *Initial adaptation.* The 'gain' (f-I slope) of a motoneuron is much higher just after the abrupt onset of a step of stimulating current than later on. Thus, motoneurons are more sensitive to *changes* of excitation than to the steady state excitation level. One of the consequences of this dynamic sensitivity is that, for all stimulation intensities except juxta-threshold ones, a motoneuron which is activated by a step of direct current tends to start off with a few high-rate intervals before it settles down to more regular firing. This *initial adaptation* probably depends, to an important extent, on the non-linear 'summation' of successive AHP-related conductance changes (Kernell & Sjöholm, 1973; Baldissera & Gustafsson, 1974; Barrett et al., 1980). The initial burst enables the motoneurons to achieve a high rate of rise of force at the sudden onset of a contraction. These aspects of motoneuronal sensitivity have been explored in detail in a series of experiments by Baldissera et al. (1987 and earlier).

5) *Late adaptation.* Following the initial adaptation there is a more gradual phase of decline in firing rate during the course of constant stimulation. This *late adaptation* is not associated with any evident gain-changes, and the drop in firing rate is particularly marked during the first 1/2-1 min of a maintained stimulation period (maximally 4 min studied; Kernell & Monster, 1982a). In one and the same neuron, the extent of late adaptation is greater at high starting rates than at lower ones. Such a relationship with discharge rate is what one would expect if the late adaptation were dependent upon cumulative after-effects of many consecutive spikes. The frequency-dependence of late adaptation is also of importance for motoneuron-muscle matching. As a result of differences in minimum rate, slow-twitch gastrocnemius motoneurons would start discharging at a lower rate than that of the fast-twitch ones if all cells were activated by the same weak suprathreshold current intensity. Consequently, under these circumstances, the slow-twitch neurons also showed a much less marked late adaptation than that of the fast-twitch cells (Kernell & Monster, 1982b). Due to their modest amount of late adaptation, the slow-twitch motoneurons are particularly well suited for the maintenance of long-lasting and steady postural contractions.

During the initial minute of a maximum voluntary contraction there is a progressive fall in motoneuronal discharge rate, similar to that of the late adaptation during constant-current stimulation. Although much of the frequency-change during a maximum voluntary contraction may be caused by reflex mechanisms (Woods et al., 1987), the late adaptation is likely to be of importance in this context as well.

4 Long-term consequences of motoneuronal activity patterns

Until now two main types of motoneuron - muscle unit match have been discussed: 1) a *recruitment-match* whereby the recruitment order of motoneurons is matched, in a functionally relevant manner, to the contractile properties of their muscle units (i.e., weak before strong, slow before fast, fatigue-resistant before fatigue-sensitive; Burke, 1981; Henneman & Mendell, 1981); 2) a *rate-match* whereby the rates of the neuronal frequency-current curve are matched by the rates of the tension-frequency curve of the muscle unit. The recruitment-match seems, to an important extent, to reflect a match between the motoneuronal membrane resistivity (and, hence, electrical excitability) and the muscle unit speed, endurance and maximum force. The rate-match reflects a match between the time courses of the motoneuronal AHP and the muscle unit twitch ('twitch-speed'). It should be noted that muscle unit *speed* comes in as a factor in both kinds of match.

How might these various kinds of motoneuron - muscle unit match be explained? To what an extent might the matching be the long-term consequence of differential muscle unit 'training', as caused by the activity patterns of various classes of motoneuron? With this type of general question in mind we embarked on a series of experiments concerning the effects of long-term stimulation on muscle properties. Preceding studies had clearly shown that a fast muscle could be made slower if it were treated continuously at a slow pulse rate (e.g., 10 Hz) during a number of weeks (Salmons & Vrbová, 1969; for other references, see reviews: Salmons & Henriksson, 1981; Pette & Vrbová, 1985). It was still rather unclear, however, what the importance were of different stimulation parameters for the type of physiological effect obtained. Does, for instance, the rate-match between motoneurons and their muscle units result because of *rate-specific* effects of activity on contractile speed?

Our chronic stimulation experiments were performed on cats whose experimental hindlimb had been made insensitive by aid of a dorsal rhizotomy (no pain, no reflex discharges). In itself, this preparatory operation (combined with an ipsilateral hemispinalization) had surprisingly small effects on the investigated muscle properties (Eerbeek et al., 1984; Donselaar et al., 1987). Our chronic stimulation patterns were given in different amounts and at different pulse rates for various groups of cats, and the total duration of treatment was 4 or 8 weeks. The daily amounts varied between 0.5 and 50% of total time, all stimulation was of supramaximal intensity, and the pulse rates were between 5 and 100 Hz. At the end of a period of chronic stimulation, the isometric contractile properties and EMG behaviour were investigated for one of the chronically activated muscles (PerL) and for its contralateral control. Thereafter, the muscles were removed and prepared for subsequent morphometric and histochemical analysis. The results led to the following main conclusions (Eerbeek et al., 1984; Donselaar et al., 1987; Kernell et al., 1987a,b):

1) *Isometric speed*. Physiologically, our studies concerned measurements of twitch time course as well as of the tension-frequency relation. Histochemically, we looked at muscle composition with respect to myosin ATPase. Our results confirmed that great daily amounts of chronic stimulation (50% of time) turns a mixed hindlimb muscle into a markedly slow one with a homogenous histochemical composition (cf. Salmons & Henriksson, 1981; Pette & Vrbová, 1985). Contrary to our own initial expectations, which were inspired by preceding publications concerning other kinds of preparation (cf. Salmons & Vrbová, 1969; Lømo et al., 1980), the effects of chronic stimulation on muscle speed were *not* dependent on the pulse rate used during the chronic treatment (cf. Figure 2). Even bursts at rates of 100 Hz could have an appreciable slowing effect on muscle contraction, and 100 Hz activation did not counteract any of the slowing effects produced by 10 Hz stimulation (for further details, see Kernell et al., 1987a). Physiologically, weak slowing effects were seen already for a stimulation pattern covering only 0.5% of daily time.

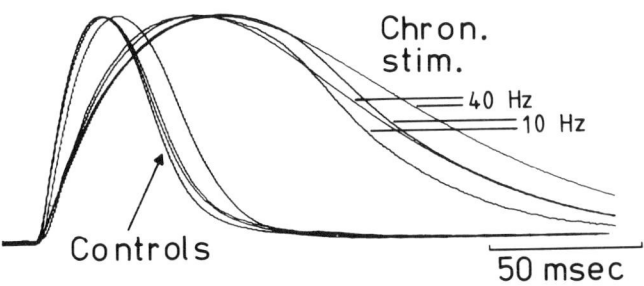

Figure 2. Records illustrating the relative unimportance of stimulus rate for long-term effects of activation on the twitch speed of cat's fast muscle (PerL). The twitches were from 8 different muscles of 4 different animals. Records labelled 'Chron.stim.' were from left-side muscles subjected to great daily amounts of long-term activation at pulse rates of 40 Hz (2 cases) and 10 Hz (2 cases) respectively. In each animal, the chronic activation covered 50% of daily time during 8 weeks. Records labelled 'Controls' were from the right-side muscles of the same animals. All records are averages of 10 sweeps each. In order to facilitate comparisons of time course, the twitches are displayed at a common time scale but with normalised amplitudes. From Eerbeek et al. (1984).

2) *Maximum tetanic force*. We confirmed that great daily time-amounts of activation tend to make a muscle considerably weaker, and that part of this weakening was caused by a decrease of fibre diameter (cf. Salmons & Henriksson, 1981; Pette & Vrbová, 1985). Furthermore, we found that the weakening as well as the shrinkage of fibre diameter were both dependent on the pulse rates used during chronic activation. For the maintenance of fibre size and maximum force, high pulse rates, producing strong contractions, tended to be more beneficial than slower rates. In mixture patterns of treatment, the weakening effects of 10 Hz activation could even be neutralised by the addition of brief 100 Hz bursts.

3) *Fatigue-resistance and EMG-behaviour*. We confirmed that great daily amounts of chronic stimulation produces a marked improvement of contractile fatigue resistance (cf. Salmons & Henriksson, 1981; Pette & Vrbová, 1985). Furthermore, we added the new observation that chronic stimulation also is effective in counteracting the 'EMG depression' that is commonly associated with contractile fatigue (i.e., the gradual decrease in amplitude of the successive compound muscle action potentials that are evoked by nerve stimulation during a fatigue test). The further analysis showed, however, that there was no strong linkage between the EMG depression and the contractile fatigue: EMG depression could be markedly affected also by daily time- amounts of stimulation (0.5%) that were too small for any pronounced change of contractile fatigue (Kernell et al., 1987b).

4) *Combined effects*. The contractile properties of a normal mixed cat muscle are typically dominated by those of relatively fatigue-sensitive units (FF and F(int); Burke, 1981). In the PerL the FF and F(int) units are together responsible for totally 77% of the maximum force (Kernell et al., 1983a). In general, chronic stimulation tended to affect the FF-F-(int)-dominated muscle such that its average properties became shifted toward those of the units which are normally most heavily used: the S units. The stimulated muscle became slower, weaker and more fatigue resistant. This shift occurred such that moderate time-amounts of daily activity caused a mild slowing and a marked improvement of fatigue resistance, i.e. a shift from an 'FF-muscle' to an 'FR-muscle'. An increase from moderate (5%) to great (50%) time-amounts of daily activation caused little further enhancement in standard measures of fatigue resistance, but gave rise to a marked decrease in speed and force, i.e. a shift from an 'FR-muscle' to an 'S-muscle'. As far as is known, the time amounts

19

required for these changes were compatible with the daily activity patterns of normal motoneurons. For instance, 'FR-muscles' were produced by daily time- amounts of activity of 5%, which is similar to the amount of spontaneous daily firing observed for presumed FR motoneurons in the rat (Hennig & Lømo, 1985; for daily activity times of whole cat muscles, see Hensbergen and Kernell, in press). Our findings are consistent with the idea that the normal matching between the properties of motoneurons and muscle units would be promoted by the differential muscle-training effects caused by motoneuronal activity patterns. However, it should be stressed that our findings do not necessarily mean that activity would be the *only* factor of importance for the differentiation of muscle unit properties. During the ontogenetic development of muscle, non-neuronal factors are known to be of great importance for differentiation (e.g., Miller & Stockdale, 1987). Such factors might, for instance, be responsible for setting the 'adaptive range' within which the adult muscle properties may become further adjusted by means of long-term effects of usage.

In the adult, usage-dependent changes would be needed for the continuous adjustment of muscle unit properties to alterations in motor behaviour. Usage-dependent changes might also be responsible for much of the remarkable recovery of motoneuron - muscle unit matching after re-innervation (cf. Gordon & Stein, 1982).

4.1 Effects of long-term activation on motoneurons

Rather little is still known about the long-term effects of usage-changes on motoneurons. Functional properties of motoneurons may be altered by spinalization (e.g. Czéh et al., 1978; Munson et al., 1986) as well as by muscle activation (Czéh et al., 1978). In general, however, the properties of motoneurons seem to be less usage-dependent than those of their muscle fibres. In our experimental situation (dorsal roots cut), the chronic nerve stimulation would cause muscles to contract and evoke antidromic discharges of their motoneurons, but there would be few direct effects on the intra-spinal synaptic activity (presumably mainly some activation of recurrent inhibition). This particular type of chronic activation had little or no effect on the respective motoneuronal cell bodies with respect to their size or oxidative enzyme activity (succinate dehydrogenase histochemistry; Donselaar et al., 1986).

Author's note. The present review is a shortened and adapted version of Kernell (1990), in which the reader may find some further details and references (see also Kernell, 1992).

References

Bakels, R. & Kernell, D. (in press). Matching between motoneurone and muscle unit properties in rat medial gastrocnemius. *Journal of Physiology*.

Baldissera, F. & Gustafsson, B. (1974). Firing behaviour of a neurone model based on the afterhyperpolarisation conductance time course and algebraical summation. Adaptation and steady state firing. *Acta Physiologica Scandinavica*, **92**, 27-47.

Baldissera, F., Campadelli, P. & Piccinelli, L. (1987). The dynamic response of cat gastrocnemius motor units investigated by ramp current injection into their motoneurones. *Journal of Physiology*, **387**, 317-330.

Barrett, E.F., Barrett, J.N. & Crill, W.E. (1980). Voltage-sensitive outward currents in cat motoneurones. *Journal of Physiology*, **304**, 251-276.

Burke, R.E., Jankowska, E. & ten Bruggencate, G. (1970). A comparison of peripheral and rubrospinal synaptic input to slow and fast twitch motor units of triceps surae. *Journal of Physiology*, **207**, 709-732.

Burke, R.E., Levine, D.N., Tsairis, P. & Zajac, F.E. (1973). Physiological types and histochemical profiles in motor units of the cat gastrocnemius. *Journal of Physiology*, **234**, 723-748.

Burke, R.E. (1981). Motor units: anatomy, physiology and functional organization. In V.B. Brooks (Ed.), *Handbook of Physiology; The Nervous System II, Part 1* (pp 345-422). Bethesda, MD: American Physiological Society.

Cooper, S. & Eccles, J.C. (1930). The isometric responses of mammalian muscles. *Journal of Physiology*, **69**, 377-385.

Cullheim, S. (1978). Relations between cell body size, axon diameter and axon conduction velocity of cat sciatic alpha-motoneurons stained with horseradish peroxidase. *Neuroscience Letters*, **8**, 17-20.

Czéh, G., Gallego, R., Kudo, N. & Kuno, M. (1978). Evidence for the maintenance of motoneurone properties by muscle activity. *Journal of Physiology*, **281**, 239-252.

Desmedt, J.E. & Godaux, E. (1981). Spinal motoneuron recruitment in man: rank deordering with direction but not with speed of voluntary movement. *Science*, **214**, 933-936.

Donselaar, Y., Kernell, D., Eerbeek, O. & Verhey, B.A. (1985). Somatotopic relations between spinal motoneurones and muscle fibres of the cat's musculus peroneus longus. *Brain Research*, **335**, 81-88.

Donselaar, Y., Kernell, D. & Eerbeek, O. (1986). Soma size and oxidative enzyme activity in normal and chronically stimulated motoneurones of the cat's spinal cord. *Brain Research*, **385**, 22-29.

Donselaar, Y., Eerbeek, O., Kernell, D. & Verhey, B.A. (1987). Fibre sizes and histochemical staining characteristics in normal and chronically stimulated fast muscle of cat. *Journal of Physiology*, **382**, 237-254.

Eerbeek, O., Kernell, D. & Verhey, B.A. (1984). Effects of fast and slow patterns of tonic long-term stimulation on contractile properties of fast muscle in the cat. *Journal of Physiology*, **352**, 73-90.

Enoka, R.M. & Stuart, D.G. (1984). Henneman's 'size principle': current issues. *Trends in Neurosciences*, **7**, 226-228.

Fleshman, J.W., Munson, J.B., Sypert, G.W. & Friedman, W.A. (1981). Rheobase, input resistance and motor-unit type in medial gastrocnemius motoneurons in the cat. *Journal of Neurophysiology*, **46**, 1326-1338.

Fleshman, J.W., Segev, I. & Burke, R.E. (1988). Electrotonic architecture of type-identified motoneurons in the cat spinal cord. *Journal of Neurophysiology*, **60**, 60-85.

Garnett, R. & Stephens, J.A. (1981). Changes in the recruitment threshold of motor units produced by cutaneous stimulation in man. *Journal of Physiology*, **311**, 463-473.

Gordon, T. & Stein, R.B. (1982). Reorganization of motor-unit properties in reinnervated muscles of the cat. *Journal of Neurophysiology*, **48**, 1175-1190.

Gustafsson, B. & Pinter, M.J. (1984). An investigation of threshold properties among cat spinal alpha-motoneurones. *Journal of Physiology*, **357**, 453-483.

Gustafsson, B. & Pinter, M.J. (1985). On factors determining orderly recruitment of motor units: a role for intrinsic membrane properties. *Trends in Neurosciences*, **8**, 431-433.

Haar Romeny, ter, B.M., Denier van der Gon, J.J. & Gielen, C.C.A.M. (1984). Relation between location of a motor unit in the human biceps brachii and its critical firing levels to different tasks. *Experimental Neurology*, **85**, 631-650.

Henneman, E. & Mendell, L.M. (1981). Functional organization of motoneuron pool and its inputs. In V.B. Brooks (Ed.), *Handbook of Physiology - The Nervous System II, Part 1* (pp. 423-507). Bethesda, MD: American Physiological Society.

Hennig, R. & Lømo, T. (1985). Firing patterns of motor units in normal rats. *Nature*, **314**, 164-166.

Hensbergen, E. & Kernell, D. (1992 a). Task-related differences in distribution of electromyographic activity within peroneus longus muscle of spontaneously moving cats. *Experimental Brain Research*, **89**: 682-685.

Hensbergen, E. & Kernell, D. (in press). Daily 'endurance-demands' in cat's ankle muscles. In A.J. Sargeant & D. Kernell (Eds.), *Neuromuscular Fatigue*. Amsterdam, Koninklijke Nederlandse Akademie van Wetenschappen.

Hoffer, J.A., Loeb, G.E., Sugano, N., Marks, W.B., O'Donovan, M.J. & Pratt, C.A. (1987). Cat hindlimb motoneurons during locomotion. III. Functional segregation in sartorius. *Journal of Neurophysiology*, **57**, 554-562.

Kandou, T.W.A. & Kernell, D. (1989). Distribution of activity within the cat's peroneus longus muscle when activated in different ways via the central nervous system. *Brain Research*, **486**, 340-350.

Kernell, D. (1965). The limits of firing frequency in cat lumbosacral motoneurones possessing different time course of afterhyperpolarisation. *Acta Physiologica Scandinavica*, **65**, 87-100.

Kernell, D. (1966). Input resistance, electrical excitability, and size of ventral horn cells in cat spinal cord. *Science*, **152**, 1637-1640.

Kernell, D. & Sjöholm, H. (1973). Repetitive impulse firing: comparisons between neurone models based on 'voltage clamp equations' and spinal motoneurones. *Acta Physiologica Scandinavica*, **87**, 40-56.

Kernell, D. & Sjöholm, H. (1975). Recruitment and firing rate modulation of motor unit tension in a small muscle of the cat's foot. *Brain Research*, **98**, 57-72.

Kernell, D. & Monster, A.W. (1981). Threshold current for repetitive impulse firing in motoneurones innervating muscle fibres of different fatigue sensitivity in the cat. *Brain Research*, **229**, 193-196.

Kernell, D. & Zwaagstra, B. (1981). Input conductance, axonal conduction velocity and cell size among hindlimb motoneurones of the cat. *Brain Research*, **204**, 311-326.

Kernell, D. & Monster, A.W. (1982 a). Time course and properties of late adaptation in spinal motoneurones in the cat. *Experimental Brain Research*, **46**, 191-196.

Kernell, D. & Monster, A.W. (1982 b). Motoneurone properties and motor fatigue. An intracellular study of gastrocnemius motoneurones of the cat. *Experimental Brain Research*, **46**, 197-204.

Kernell, D., Eerbeek, O. & Verhey, B.A. (1983 a). Motor unit categorization on basis of contractile properties: an experimental analysis of the composition of the cat's m. peroneus longus. *Experimental Brain Research*, **50**, 211-219.

Kernell, D., Eerbeek, O. & Verhey, B.A. (1983 b). Relation between isometric force and stimulus rate in cat's hindlimb motor units of different twitch contraction time. *Experimental Brain Research*, **50**, 220-227.

Kernell, D. (1984). The meaning of discharge rate: excitation-to-frequency transduction as studied in spinal motoneurones. *Archives Italiennes de Biologie*, **122**, 5-15.

Kernell, D., Verhey, B.A. & Eerbeek, O. (1985). Neuronal and muscle unit properties at different rostro-caudal levels of cat's motoneurone pool. *Brain Research*, **335**, 71-79.

Kernell, D., Eerbeek, O., Verhey, B.A. & Donselaar, Y. (1987 a). Effects of physiological amounts of high- and low-rate chronic stimulation on fast-twitch muscle of the cat hindlimb. 1. Speed- and force-related properties. *Journal of Neurophysiology*, **58**, 598-613.

Kernell, D., Donselaar, Y. & Eerbeek, O. (1987 b). Effects of physiological amounts of high- and low-rate chronic stimulation on fast-twitch muscle of the cat hindlimb. II. Endurance-related properties. *Journal of Neurophysiology*, **58**, 614-627.

Kernell, D. & Zwaagstra, B. (1989). Dendrites of cat's spinal motoneurones: relationship between stem diameter and predicted input conductance. *Journal of Physiology*, **413**, 255-269.

Kernell, D. (1990). Spinal motoneurons and their muscle fibers: mechanisms and long-term consequences of common activation patterns. In M.D. Binder & L.M. Mendell (Eds.), *The Segmental Motor System* (pp. 36-57). New York, Oxford University Press.

Kernell, D. (1992). Organized variability in the neuromuscular system: A survey of task-related adaptations. *Archives Italaliennes Biologie*, **130**, 19-66.

Kugelberg, E. & Thornell, L.-E. (1983). Conduction time, histochemical type, and terminal cisternae volume of rat motor units. *Muscle & Nerve*, **6**, 149-153.

Lømo, T., Westgaard, R.H. & Engebretsen, L. (1980). Different stimulation patterns affect contractile properties of denervated rat soleus muscles. In D. Pette (Ed.), *Plasticity of Muscle* (pp.297-309). Berlin, Walter de Gruyter & Co.

Lüscher, H.-R., Ruenzel, P. & Henneman, E. (1979). How the size of motoneurones determines their susceptibility to discharge. *Nature*, **282**, 859-861.

Miller, J.B. & Stockdale, F.E. (1987). What muscle cells know that nerves don't tell them. *Trends in Neurosciences*, **10**, 325-329.

Munson, J.B., Foehring, R.C., Lofton, S.A., Zengel, J.E. & Sypert, G.W. (1986). Plasticity of medial gastrocnemius motor units following cordotomy in the cat. *Journal of Neurophysiology*, **55**, 619-634.

Pette, D. & Vrbová, G. (1985). Neural control of phenotypic expression in mammalian muscle fibres. *Muscle & Nerve*, **8**, 676-689.

Pinter, M.J., Curtis, R.L. & Hosko, M.J. (1983). Voltage threshold and excitability among variously sized cat hindlimb motoneurons. *Journal of Neurophysiology*, **50**, 644-657.

Salmons, S. & Vrbová, G. (1969). The influence of activity on some contractile characteristics of mammalian fast and slow muscles. *Journal of Physiology*, **201**, 535-549.

Salmons, S. & Henriksson. J. (1981). The adaptive response of skeletal muscle to increased use. *Muscle & Nerve*, **4**, 94-105.

Swett, J.E., Eldred, E. & Buchwald, J.S. (1970). Somatotopic cord-to- muscle relations in efferent innervation of cat gastrocnemius. *American Journal of Physiology*, **219**, 762-766.

Ulfhake, B. & Kellerth, J.-O. (1981). A quantitative light microscopic study of the dendrites of cat spinal alpha-motoneurons after intracellular staining with horseradish peroxidase. *Journal of Comparative Neurology*, **202**, 571-583.

Ulfhake, B. & Kellerth, J.-O. (1982). Does alpha-motoneurone size correlate with motor unit type in cat triceps surae? *Brain Research*, **251**, 201-209.

Ulfhake, B. & Kellerth, J.-O. (1984). Electrophysiological and morphological measurements in cat gastrocnemius and soleus alpha-motoneurones. *Brain Research*, **307**, 167-179.

Woods, J.J., Furbush, F. & Bigland-Ritchie, B. (1987). Evidence for a fatigue-induced reflex inhibition of motoneuron firing rates. *Journal of Neurophysiology*, **58**, 125-137.

Zwaagstra, B. & Kernell, D. (1981). Sizes of soma and stem dendrites in intracellularly labelled alpha-motoneurones of the cat. *Brain Research*, **204**, 295-309.

22

Chapter 3: Properties of submaximally stimulated muscle and some functional consequences of gradation of muscle force

P.A. Huijing

1 Introduction

At the symposium and in this volume Daniel Kernell presented a very nice review of mechanisms of force gradation, i.e., recruitment gradation and rate gradation. In addition, he presented a review of long term effects of motoneural activity patterns on motor unit properties. During the discussion following his lecture a number of issues were raised regarding immediate functional consequences of the use of recruitment and rate gradation of force. A number of such consequences will be treated below. It should be noted that the discussion is limited to muscles in the isometric condition. Some methodological aspects of electrical stimulation of muscles through their peripheral nerves will be discussed (which also reflects part of the discussion at the symposium). Subsequently, effects on length force characteristics will be considered.

2 Experimental submaximal stimulation of muscle

In myology research characteristics of active muscles are most frequently studied under conditions of maximal activation of the muscle (e.g., Hill, 1970; Close, 1972). A major reason for this can be found in the fact that experimentally such activation is easily obtained by supramaximal electrical stimulation of the periferal nerve (i.e., through external electrodes) and that the force response thus obtained is highly reproducible. Most researchers are very well aware of the fact that compared to the real life situation very unnatural conditions are created: All motor units are active at unusually high firing rates in a synchronised fashion.

Despite these facts, in movement sciences, where research is very often inspired by a desire to obtain information which is useful in interpreting phenomena of real movement, this form of stimulation is well accepted because it allows probing into neuromuscular mechanisms on the basis of scientifically sound procedures. A lot of information about basic properties of neuromuscular complexes has been obtained this way and this information is used to make statements about natural movement by inference or modelling.

Using the same extraneural stimulation electrode set up, the simplest way to go from maximally to submaximally activated muscle would be to reduce the current used for stimulation of the nerve. As a consequence, some motor units that were active initially will be derecruited as the applied current is not sufficient anymore to excite their axons. Electrical stimulation of a nerve by extracellular electrodes will lead to excitation of large motoneurons before smaller ones (Blair & Erlanger, 1933). As a consequence the smaller motor units are first to be derecruited in the described situation of stimulation, followed by derecruitment of bigger units as the current is decreased more. It is obvious that this is in contrast with the size principle of recruitment which is active if motor units are 'turned on or off' through variation of synaptic excitation (e.g., Henneman et al., 1965; Milner Brown et al., 1974).

Unpublished observations (Huijing & Koper) indicate that the force response is not very reproducible when the experiment was performed this way, which may be related to small variations of the position of the nerve with respect to the electrodes or other causes. These types of artefacts make adequate application of these techniques difficult.

It should be noted that at Twente University work is in progress that is directed at the design of intrafascicular multipolar electrodes which would allow more control over

Author's address: Department of Functional Anatomy, Faculty of Human Movement Sciences, Vrije Universiteit, Van der Boechorststraat 9, 1081 BT Amsterdam, The Netherlands.

recruitment order of individual motor units. Modelling indicates that appropriately choosing the distance of poles of intrafascicular electrodes in relation to the internodal distance of nodes of Ranvier will allow size principle recruitment order (Meier et al., 1992). It is concluded that the best stimulation characteristics should be obtained with an intraneural multi-electrode that contains a large number of electrodes sites. Of this electrode each site is expected to selectively stimulate a small group of motor units or individual motor units (Meier, 1992). Experimentally such electrodes have been produced using silicon technology (Meier, 1992). As it is expected that further development of this technique will require a substantial amount of time before such electrodes will be generally available, it is of interest to look into the effects of more conventional techniques of obtaining size principle ordered recruitment using extraneural electrical stimulation.

In the last century (Wedensky, 1884) it became evident that conduction of action potentials along a nerve could be 'blocked' by applying a high frequency stimulation. In the last decades these methods have been reapplied to study effects of dynamic recruitment and derecruitment of motor units with simultaneous variation of firing frequency (Solomonow et al., 1983; Solomonow, 1984; Zhou et al., 1987; Barrata et al., 1989; Baratta & Solomonow, 1991). This technique involves classical stimulation of the nerve in combination with simultaneous high frequency stimulation (600 Hz) at a more distal location. Using this technique, recruitment and derecruitment of motor units is accomplished according to size (Tanner, 1962; Muller & Hunsperger, 1967), i.e., large size motor units are recruited last and derecruited first, the order of which is comparable to recruitment and derecruitment obtained by natural variation of synaptic input (Henneman et al., 1965). A drawback of this method is that the exact mechanism by which the blocking of stimulation of motor nerves occurs is not known. Two types of hypotheses can be distinguished in the literature:
1) Anodal blocking conditions at the distal set of electrodes prevent conduction of action potentials on the motor axons by hyperpolarising the transmembrane potential (e.g., Fang & Mortimer, 1991). This would be expected to lead to a sudden arrest of firing of the motor unit.
2) The 600 Hz stimulation prevents conduction of action potentials from nerve to muscle (possibly by depletion of neurotransmitter substance in the motor endplates) (e.g., Zhou et al., 1987). In this case a progressive decrease of motor unit firing rate until derecruitment would be expected. It is even conceivable that both mechanisms act simultaneously.
At present the possibility can not be excluded that the high frequency stimulation may interfere with processes of conduction of the action potential along the muscle fibres or into the t-tubules, or with processes of excitation contraction coupling. In most of such cases a gradual decrease of firing frequency of muscle fibres would not be unlikely.

3 Different mechanical properties of motor units

If motor units differed only in size and thus only in maximal force they can produce, selective activation of specific motor units would obviously only result in a grading of the force output of the muscle without altering the muscle's other mechanical characteristics. As motor units of different size have different mechanical characteristics, recruitment of specific motor units may lead to altered mechanical properties of the muscle depending on the level of excitation of the motor unit pool. We will consider effects on muscle length force characteristics below.

3.1 Distribution of sarcomere and fibre optimum lengths with respect to muscle optimum length

The length of sarcomeres within a muscle fibre is a major determinant for the maximal amount of force that can be exerted by that fibre (sarcomere length force curve). Very often muscle is thought to be build of fibres which may differ with respect to certain mechanical properties (e.g., speed of contraction, resistance to fatigue, fibre diameter, and number of sarcomeres in series) but is uniform with respect to fibre length and/or sarcomere length of different fibres (e.g., Rack & Westbury 1969; Nordström & Yemm, 1974; Tardieu et al., 1977; Lieber & Boakes, 1988). Particularly in muscle modelling the assumption of uniform architecture with respect to this is almost invariably made (e.g., Gans, 1982; Huijing & Woittiez, 1984; Woittiez & Huijing, 1984; Otten, 1988; Bobbert et al., 1986; Zajac, 1989; van Leeuwen & Spoor, 1992; Zuurbier & Huijing, 1992). However more and more evidence is accumulating indicating that the mean sarcomere lengths within different fibres of a muscle is not uniform but may show a certain distribution. This means that at a given muscle length, lengths of sarcomeres within different fibres may be quite different. Another way of illustrating this phenomenon is to consider the fact that optimum sarcomere or fibre length is distributed with respect to muscle optimum length. This principle is illustrated in Figure 1.

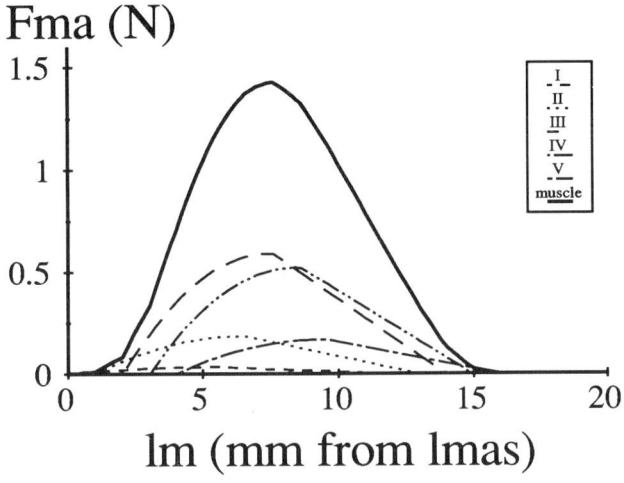

Figure 1. Length force characteristics of a purely hypothetical parallel fibred modelled muscle consisting of five (groups of) motor units, whose optimum length is distributed with respect to muscle optimum length. At any muscle length the muscle force is calculated by the sum of forces of individual (groups of) motor units at that length. Length force curves are shown for the (groups of) motor units as well as for the muscle. Length is expressed relative to muscle active slack length (lmas, i.e., the lowest length at which any active force is generated).

For distal fibres of rat medial gastrocnemius muscle (GM) differences between tetanus fibre optimum length and fibre length at tetanus muscle optimum length of approximately 1 mm were found (Heslinga & Huijing, 1990). From the results of Grottel et al. (1990) it can be deduced that in that muscle, muscle length range of twitch optimum force exertion for a set of motor units studied was approximately 2 mm. These values represent approximately 12.5 - 25 % of the length range between muscle optimum and active slack length. Therefore it may be concluded that this feature may be a rather important factor for the determination of the length range of active force exertion of rat muscle. This conclusion is strengthened by evidence that aspects of the distribution may be prone to adaptation to externally imposed conditions (i.e., long term effect).

For cat muscle evidence for the existence of such distribution of optimum lengths may be found in the work of Lewis et al. (1972) (flexor hallucis longus), Bagust et al. (1973) (flexor digitorum longus), and Stephens et al. (1978) (GM).

If big and small motor units would be randomly distributed over groups of motor units having their optimum force at a given length, the effects on muscle length force characteristics at different levels of excitation would be limited. However some evidence is available that this is not the case in some muscles at least. For cat flexor digitorum muscle (Bagust et al., 1973) it was found that slow motor units attained their optimum force at higher muscle lengths. Results for rat GM (Grottel et al., 1990), indicated a similar relationship between motor unit size and twitch optimum length. Huijing and Baan (1992) recently reported a comparison between length force characteristics at maximal and submaximal levels of tetanic activation of rat GM (Figure 2). These results showed that tetanus optimum length was encountered at higher muscle length as the muscle activation was decreased by high frequency stimulation of the nerve in addition to supramaximal 100 Hz stimulation. During this decreasing activation it is likely that motor units of decreasing size are derecruited. The size of this shift was relatively large, as it was equal to approximately 50% of the length range between optimum and active slack of the fully activated muscle. It was at least in part ascribed to the presence of the distribution of optimum lengths. As Daniel Kernell also pointed out during the discussion after his presentation at the symposium, presently the evidence for a nonrandom appropiation of motor units of different size to certain sub-populations of the distributed motor unit optimum lengths population is too limited to make sweeping statements about muscle in general. However the potential functional significance of such a feature warrants further experimental work in this area.

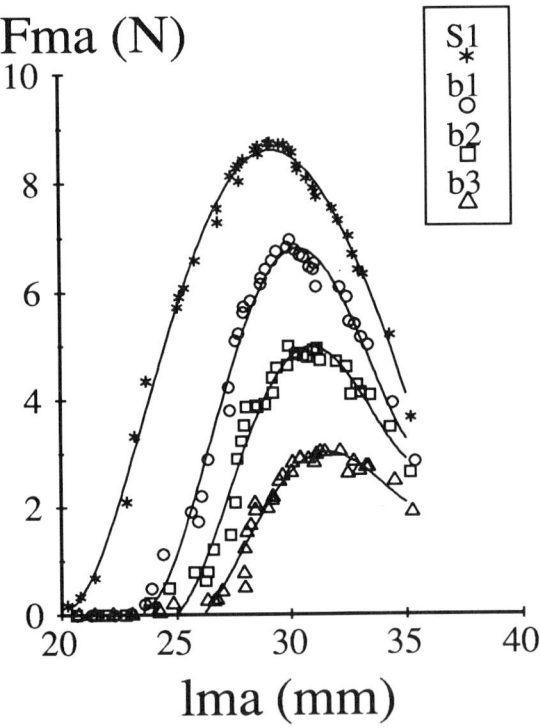

Figure 2. Length force characteristics of rat medial gastrocnemius muscle in maximally activated muscle (s1) and at different submaximal levels of activation (b1-b3) obtained by simultaneous supramaximal stimulation of the nerve at 100 Hz in combination with stimulation at 600 Hz for which the current was adjusted for optimal effect. Typical results for raw data of one muscle are shown in combination with polynomial curves fitted to it. The length at which the maximum of each fitted curve was encountered was defined as optimum length for that condition. Note that at lower levels of activation muscle optimum length shifts to higher values.

3.2 Effects of firing frequency on motor unit and muscle characteristics

In his paragraph on rate gradation Dr. Kernell introduced the sigmoid tension frequency curve of motor units. Ample evidence is available that specific characteristics of the force- and tension-frequency curves are different for fast and slow motor units (for references see Dr. Kernell's presentation, this volume). It is pointed out in Dr. Kernell's contribution that differences of this relation between slow (i.e., smaller) units and fast (bigger) ones are predominantly related to kinetics of intracellular calcium movements.

One effect on length force characteristics is immediately apparent from force frequency curves: Force exerted by motor units is dependent on firing frequency and this dependency is used optimally in rate coding gradation of force (Kernell, this volume).

An effect of firing frequency on muscle characteristics not immediately apparent from frequency force curves can be illustrated by the properties of muscle maximally stimulated at different firing rates. If all motor units of a muscle are induced to fire at the same rate, variation of that rate results in an effect on muscle optimum length, as shown by Rack and Westbury (1969) for cat soleus muscle: At lower firing rates optimum length is found at higher muscle length. This finding was recently confirmed for rat GM in our laboratory (Roszek et al., 1992). Figure 3 shows a preliminary result of this analysis.

Rack and Westbury (1969) hypothesised that these effects were related to length dependent changes of activation (thus presumably calcium dependent events). However results of Rome et al. (1985) indicate (at least for frog muscle fibres) that optimum length of isolated fibres is unchanged by lowering firing frequency but that for a give sarcomere length below optimum *normalised* force was less at lower firing frequencies. If these results may be generalised to mammalian muscle, it would mean that any changes in length range of active force generation could at least in part be ascribed to intracellular events but that explanations for shifting muscle optimum length should be found at the level of muscular organ level (i.e., possibly distribution effects as described above). It is clear that not sufficient material is available presently to draw such conclusions unequivocally.

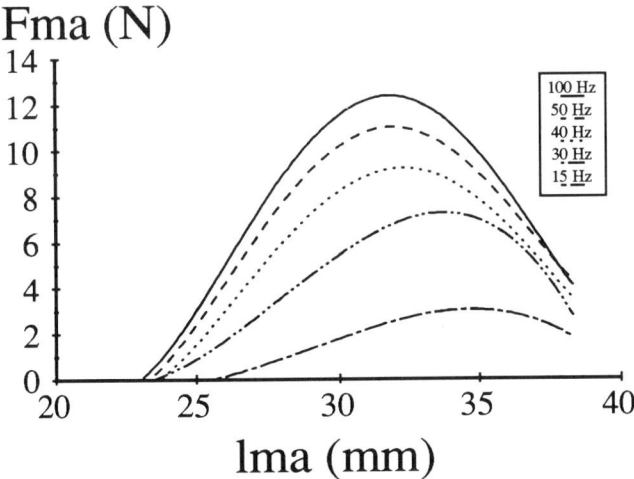

Figure 3. Length force characteristics of rat gastrocnemius medialis muscle obtained during supramaximal excitation of the nerve at different frequencies. Curves fitted to experimental data of one muscle are shown, these results were representative for the group results of the muscles studied. The length at which the maximum of each fitted curve was encountered was defined as optimum length for that condition. Note that at lower stimulation frequencies optimum length shifts to higher values.

4 Conclusion

It is evident that making use of the two main mechanisms of force gradation may lead to variable mechanical properties of the partially activated muscle. For that reason generalisation of results for maximally activated muscle to conditions of real life movement through inference or modelling should be handled with utmost care. Experimentation with maximally activated muscle should continue as it clearly helps to increase understanding of the neuromuscular processes. However more attention should be directed towards increasing the amount and quality of experimental work aimed at studying mechanical and other properties of submaximally active muscles.

References

Bagust, J., Knott, S., Lewis, D.M., Luck, J.C., & Westerman, R.A. (1973). Isometric contractions of motor units in a fast twitch muscle of the cat. *Journal of Physiology*, **231**, 87-104.

Barrata, R.V., Ichie, M., Hwang, S., & Solomonow, M. (1989). A method for studying muscle properties under orderly stimulated motor units. *Journal of Biomedical Engeneering*, **11**, 141-147.

Baratta, R.V., & Solomonow, M. (1991). The effect of tendon viscoelastic stiffness on the dynamic performance of isometric muscle. *Journal of Biomechanics*, **24**, 109-116.

Blair, E., & Erlanger, J. (1933). A comparison of the characteristics of axons through their individual electric responses. *The American Journal of Physiology*, **106**, 524-564.

Bobbert, M.F., Huijing, P.A., & van Ingen Schenau, G.J. (1986). A model of human triceps surae muscle-tendon complex applied to jumping. *Journal of Biomechanics*, **19**, 887-898.

Close, R.I. (1972). Dynamic properties of skeletal muscle. *Physiological Reviews*, **52**, 129-197.

Fang, Z.P., & Mortimer, J.T. (1991). A method to effect physiological recruitment order in electrically activated muscle. *IEEE Transactions on Bio-medical Engeneering*, **38**, 175-179.

Gans, K. (1982). Fiber architecture and muscle function. *Exercise and Sport Sciences Reviews*, **10**, 160-207.

Grottel, K., Celowski, J., & Anissimova, N. (1990). The course of motor unit twitch in dependence on muscle stretching force: Studies on medial gastrocnemius of the rat. *Acta Neurobiologiae Experimentalis*, **50**, 569-600.

Henneman, E., Somjen, G., & Carpenter, D.O. (1965). Functional significance of cell size. *Journal of Neurophysiology*, **28**, 560-580.

Hill, A.V. (1970). *First and last experiments in muscle mechanics*. Cambridge: Cambridge University Press.

Heslinga, J.W., & Huijing, P.A. (1990). Effects of growth on architecture and functional characteristics of adult rat gastrocnemius muscle. *Journal of Morphology*, **206**, 119-132.

Huijing, P.A., & Baan, G.C. (1992). Stimulation level-dependent length-force and architectural characteristics of rat gastrocnemicus muscle. *Journal of Electromyography and Kinesiology*, **2**, 112-120.

Huijing, P.A., & Woittiez, R.D. (1984) The effect of architecture on skeletal muscle performance: A simple planimetric model. *Netherlands Journal of Zoology*, **34**, 21-32.

Lewis, D.M., Luck J.C., & Knott, S. (1972). A comparison of isometric contractions of the whole muscle with those of motor units in a fast twitch muscle of the cat. *Experimantal Neurology*, **37**, 68-85.

Leeuwen, J.L. van, & Spoor, C.W. (1992). Modelling mechanically stable muscle architecture. *Philosophical Transactions of the Royal Society of London*, **B336**, 275-292.

Lieber, R.L., & Boakes, J.L. (1988). Sarcomere length and joint kinematics during torque production in frog hindlimb. *The American Journal of Physiology*, **254**, C759-C768.

Meier, J.H. (1992). Selectivity and design of neuro-electronic interfaces. Doctoral Dissertation University of Twente, Enschede, The Netherlands.

Meier, J.H., Rutten, W.L.C., Zoutman, A.E., Boom, H.B.K., & Bergveld, P. (1992). Simulation of multipolar fibre selective neural stimulation using intrafascicular electrodes. *IEEE Transactions on Bio-medical Engeneering*, **39**, in press.

Milner-Brown, H.S., Stein, R.B., & Yemm, R. (1974). The orderly recruitment of human motor units during voluntary isometric contractions. *Journal of Physiology*, **230**, 359-370.

Muller, A., & Hunsperger, B. (1967). Reversibele Blockierung der Erregungsleitung in nerven durch mittelfrequenz-daurestrom. *Helvetica Physiologica et Pharmacologica Acta*, **25**, CR211-CR213. (Cited from Solomonow, 1984).

Nordström, S.H., & Yemm, R. (1974). The relationship between jaw position and isometric active tension produced by direct stimulation of rat masseter muscle. *Archives of Oral Biololgy*, **419**, 353-359.

Otten, E. (1988). Concepts and models of functional architecture in skeletal muscle. *Excercise and Sport Sciences Reviews*, **16**, 89-137.

Rack, P.M.H., Westbury, D.R. (1969). The effects of length and stimulus rate on tension in the isometric cat soleus muscle. *Journal of Physiology*, **204**, 443-460.

Rome, L.C., Morgan, D.L., & Julian, F.J. (1985). Stimulation rate potentiators and sarcomere length-tension relationship of muscle. *The American Journal of Physiology*, **249** (*Cell Physiology*, **18**), C497-C502.

Roszek, B., Baan, G.C., & Huijing, P.A. (unpublished observations, 1992). Effects of stimulus frequency on length force characteristics of fully active rat muscle.

Solomonow, M. (1984). External control of the neuromuscular system. *IEEE Transactions on Bio-medical Engeneering*, **31**, 752-63.

Solomonow, M., Eldred, E., Lyman, J., & Foster, J. (1983). Control of muscle contractile force through indirect high frequency stimulation. *American Journal of Physical Medicine*, **62**, 71-82.

Stephens J.A., Reinking, R.M., & Stuart, D.G. (1978). The motor units of cat medial gastrocnemius: Electrical and mechanical properties as a function of muscle length. *Journal of Morphology*, **146**, 495-512.

Tanner, J. (1962). Reversible blocking of nerve conduction by alternating current excitation. *Nature*, **195**, 712-713.

Tardieu, C., Tabary, J.C., Huet de la Tour, E., Tabary, C., & Tardieu, G. (1977). The relationship between sarcomere length in the soleus and tibialis anterior and the articular angle of the tibia-calcaneum in cats during growth. *Jornal of Anatomy*, **124**, 581-588.

Wedensky, N. (1884). Wie rasch ermudet die Nerv? *Zentralblatt für Medizinische Wissenschaft*, 65-68. (Cited from Solomonow, 1984).

Woittiez, R.D., & Huijing, P.A. (1984). A three-dimensional muscle model: A quantified relation between form and function of skeletal muscles. *Journal of Morphology*, **182**, 95-113.

Zajac, F.E. (1989). Muscle and tendon: Properties, models, scaling, and application to biomechanics and motor control. *CRC Crit.cal Reviews in Biomedical Engeneering*, **17**, 359-411.

Zhou, B.H., Baratta, R., & Solomonow, M. (1987). Manipulation of muscle force with various firing rate and recruitment control strategies. *IEEE Transactions on Bio-medical Engeneering*, **34**, 128-139.

Zuurbier, C.J., & Huijing, P.A. (1992). Influence of muscle geometry on shortening speed of fibre, aponeurosis and muscle. *Journal of Biomechanics*, **25**, 1017-1026.

Chapter 4: The integrative nervous system: Sherrington revisited

B. Hopkins & A. Gramsbergen

1 Introduction

The birth of modern neurophysiology can be traced to the pioneering studies of two men: Charles S. Sherrington (1857-1952) and Santiago Ramón y Cajal (1852-1934). Sherrington put forward his concepts of the reflex arc and the synapse (derived from the Greek word 'to clasp tightly') towards the end of the 19th century at a time when many neuro-anatomists still held to the view that the nervous system consisted of unbroken protoplasmic connections between cellular material. The Spanish anatomist provided histological evidence that supported Sherrington's ideas on the basic functional units of the nervous system - a provision that eventuated in the neuron or cellular contact theory (DeFelipe & Jones, 1988). Together the neuron theory of Ramón y Cajal and the synaptic theory of Sherrington rejuvenated the long-standing issue of how the nervous system and its various structures controlled and regulated action (i.e., purposeful, goal-directed movements). For Sherrington, the key to promoting further understanding about the control of action was embodied in his notion of integration.

In 1904 Sherrington delivered his Silliman Lectures in which he introduced the term integration into neurology. Later published as 'The Integrative Action of the Nervous System' (IANS), Sherrington (1906) treated the term as meaning the interaction of excitatory and inhibitory pathways in producing purposeful action. At that time, he was keen to demonstrate that the reflex arc was the basic building block of the central nervous system. This stance ultimately led to Sherrington being classified as a reflexologist. However, we find in the IANS his well-known admonition that the simple reflex "... is probably a purely abstract conception because all parts of the nervous system are connected together and no part is probably ever capable of reaction without affecting and being affected by various other parts" (p. 8). Never one to reify his concepts, Sherrington eventually rejected the notion of reflexes as basic units of the central nervous system (CNS) in the preface of the 1947 reprint of the IANS.

In accordance with his view of the CNS as an organ of integration, Sherrington devoted much of his earlier experimental work to discovering the properties of the 'final common pathway'. This pathway he took to be a single motorneuron whose membrane was capable of integrating a variety of inputs. In doing so, Sherrington assiduously avoided the prevalent tendency during his day of locating reflexes in the spinal cord and voluntary movements in the motor cortex, preferring to designate the latter the 'cord area'. For Sherrington, the motor cortex was neither the main organ for controlling the output of the spinal cord nor the primary source of movement generation. For him, the integrative functions of the nervous system resided at both cortical and spinal levels. The neurophysiological condrum then was how these two distinct structures cooperated in producing purposeful actions. The best conjecture he could offer was that the motor cortex and the motor region of the spinal cord (i.e., the motorneurons in the ventral horns) were in some way tightly coupled. He did so on the basis of experiments with decorticated dogs in which he noted that destruction of the motor cortex led to degeneration of the pyramidal tract. It was not until some 50 years later that Sherrington's conjecture could receive firm neuroanatomical support due to breakthroughs in anterograde and retrograde techniques for tracing pathways and locating origin cells. Important in this respect were the pioneering studies of Kuypers (1962; 1964; 1982) which have been so ably discussed by Voogd (this volume).

Authors' address: Faculty of Human Movement Sciences, Vrije Universiteit, Van der Boechorststraat 9, 1081 BT Amsterdam, The Netherlands.

It is important to recognise that Sherrington, despite all his talk about integration, was a confirmed dualist. His Cartesian fealty was expressed in two ways. Firstly, unlike Pavlov (and other behaviourists), he portrayed the CNS as a system for adapting the animal to the external environment. As such, the brain imposed his order and organisation on the environment in a more-or-less unidirectional manner. Secondly, by the time he had written "The Brain and its Mechanisms" (1933), he was convinced that mental experience and cerebral activity had no immediate connection - the two simply coinciding in time and space. Thus, a legacy of Sherrington's endeavours to understand the workings of his 'enchanted loom' was a subsequent focus on simple, rather than complex, actions which were to be regarded as motor phenomena rather than the products of both perceptual and motor systems. Nevertheless, Sherrington's notion of integration did constitute a beginning in understanding how the perceptual and motor systems cooperate in generating various modes of action. In the present contribution we endeavour to relate this notion to what is currently known about the control of complex actions with reference to cortical structures other than the primary motor cortex. And as a further complement to Voogd's chapter, we ask "what neural substrates are implicated in the sensorimotor integration required for such actions?" Put in more fashionable terminology, can we identify particular neuroanatomical structures which are involved in perception-action coupling? In asking this question, we do not wish to imply that perception meets action in some specific region of the brain. Rather, and in keeping with Sherrington's anti-locationist attitude, we hold that perception and action are yoked together in widely distributed neural networks at both spinal and supraspinal levels. In order to illustrate the distributed nature of the potential meeting places for perception and action we range from the parietal cortex to the premotor and supplementary motor cortices of the frontal lobe, and finally to the spinal cord.

Using cytoarchitectural criteria, the cortex has been compartmentalised into 20 areas (Campbell), 47 areas (Brodmann), 109 areas (Von Economo) and more than 200 areas (Vogt). In the subsequent sections we will rely on Brodmann's classification as that is the one most commonly used in textbooks on neuroanatomy. It should be borne in mind, however, that while many of the areas in this and other classification systems can be identified by microscopic criteria, some can only be distinguished by more subtle conventions that are open to debate.

2 Posterior parietal (association) cortex

This part of the cerebrum is divided into superior (area 5) and inferior (area 7) parietal lobules. The anatomical connections of the posterior parietal cortex have been shown to be very rich and varied involving other cortical areas and a large number of subcortical structures. Cortical connections include, for example, the prestriate visual cortex, the primary somatosensory cortex, the premotor cortex and the lateral prefrontal cortex. Areas 5 and 7 share a number of connections while at the same time showing some differences. Thus, for example, area 5 receives afferents from the primary somatosensory cortex and the premotor cortex while area 7 is the recipient of cortical afferents from area 5 and the prefrontal cortex (Jones & Powell, 1970). In terms of projections, there are also differences: more pathways projecting to the temporal lobe originate in area 7 (Pandya & Kuypers, 1969). While a direct cortico-cortical connection from area 5 to the primary motor cortex (area 4) has been found (Strick & Kim, 1978), it has not yet been possible to activate this pathway by means of electrical stimulation (Seal, 1989). However, there are at least three indirect pathways by which area 5 could act on motor output. One is the cerebro-ponto-cerebellar route, another is via the premotor cortex and a third is the cortico-striate projection involving the basal ganglia.

During the last 20 years much has been discovered about the anatomical connections of the posterior parietal cortex. We are now confronted with a very complex picture which mandates care in drawing any hard and fast conclusions about its functional organisation.

In the past the posterior parietal cortex was considered to be a purely sensory region, more or less continuous in function with the somatosensory cortex, but particularly important in the maintenance of attention. This view was dramatically changed by the program of research initiated by Mountcastle et al. (1975). In this study, single cell recordings in behaving monkeys revealed discharges in area 5 cells prior to the onset of a reaching movement and even before muscle contraction. This finding was recently replicated and extended in a more sophisticated study which not only distinguished reaction from movement time but also provided a stimulus indicating the direction of the impending movement (Crammond & Kalaska, 1989). Together they lead these authors to the conclusion that area 5 is involved in the initiation of arm movements, perhaps in terms of direction or covarying parameters such as changes in joint angle. However, given the fact that the direct pathway from area 5 to area 4 has not been shown to be functional, it may be questioned as to whether area 5 really performs this role. While not necessarily involved in movement initiation, it is clear from these and other related experiments (e.g., Kalaska et al., 1983) that area 5 is not simply sensory or motor, but rather a site of sensorimotor integration once movements have been initiated.

While area 5 may be considered to form part of a sensorimotor interface, it does share some functional similarities with area 4. For example, it has been found that proximal-arm related neurons in area 5 respond with continuously graded discharges to unloaded, two-dimensional arm movements which are strikingly similar to those obtained from the primary motor cortex under the same task conditions (Georgopoulos et al., 1982). More specifically, the vector-averaged activity of populations of neurons in both areas correlated closely with the direction of reaching rather than with the starting or final positions of the movement. Clearly these findings have implications for the mass-spring model of Fel'dman (1966; 1986) in which muscles are depicted as a spring with a mass attached. In this model the CNS controls movements through a process which requires it only to specify the final position with the details of the trajectory being determined by the inherent inertial and visco-elastic properties of the limb and feedback from muscles around a joint. However, there appear to be functional differences between areas 5 and 4 which also challenge the elegant parsimony of the mass-spring model. These differences came to light when the question was posed as to whether the two areas control arm movements in two different coordinate frameworks - one being kinematic, involving the spatial parameters of position, direction, and displacement and the other dynamic in that it concerns the forces, torques, and muscle activity producing movement. In short, it was found in one study that during loaded reaching movements, the activity of most area 5 neurons was hardly altered suggesting that this part of the parietal cortex controls the spatial parameters (Kalaska et al., 1990). That the motor cortex controls the movement dynamics was adduced previously from the fact that many neurons in area 4 showed large changes in activity during loading (Kalaska et al., 1989). In both studies, the monkeys had to perform movements with little variability in velocity or acceleration. Thus, it remains open as to whether area 5 is implicated not only in the spatial control of goal-directed arm movements but also in their temporal parameterisation.

Studies on the functional properties of area 7 cells have been less frequent in recent years. Evidence to date clearly points to classes of neurons in this area possessing oculomotor functions such as fixation, pursuit, and scanning (Lynch, 1980). More recently, it has been suggested that neurons in the posterior bank of the intraparietal sulcus found in area 7 are concerned with the visual guidance of hand movements (Taira et al., 1990). More specifically, it is held that these neurons play an important role in matching efferent output to the spatial properties of the object to be manipulated. Once again we find the posterior parietal cortex being assigned a sensorimotor function, but this time relative to the control of the distal musculature. Thus, there appears to be a moderate degree of regional specificity of function in this part of the cerebrum. However, we should treat this surmisal with some caution for if anything is known about the CNS, then it is that a given neural substrate will take on a different functional role under different circumstances; That is, it will demonstrate the property of multifunctionality. For the present purposes, we can conclude that the posterior parietal cortex as a whole may be regarded as a site where task-specific couplings between perception and movement are achieved in the service of goal-directed actions such as reaching and grasping once they have been initiated. We consider next where such actions may be initiated.

3 Premotor and supplementary cortices

These two parts of the cerebrum are collectively treated as forming the nonprimary motor cortex. The premotor cortex (PMC) has been traditionally assigned to areas 6 and 8 on the lateral surface of the hemisphere and the supplementary motor cortex (SMC) to the medial surface of area 6. However, no clear cytoarchitectural differences have been found and therefore distinctions between them are better defined in terms of their neurophysiological properties.

The PMC is located immediately rostral to the primary motor cortex and immediately lateral to the SMC. The efferent and afferent connectivities of the PMC have been studied less frequently than those of other cortical areas. Furthermore, the involvement of the PMC in the control of movement has been largely ignored, probably due to earlier findings that its ablation resulted in motor deficits that were difficult to describe and that stimulating it electrically failed to elicit muscle contractions. With improvements in electrophysiological recordings in behaving monkeys and the use of positron emission tomography (PET) to measure regional cerebral blood flow in human subjects, it is becoming clear that the PMC plays a number of important roles in motor control. Such is the diversity of functions currently accorded to this motor field that it is only possible to mention a few of them here.

That the PMC can be treated as a motor field is revealed by its afferent and efferent connections. It receives inputs from the cerebellum via the thalamus (Schell & Strick, 1984) and projects onto brainstem neurons giving rise to the dorsolateral and ventromedial descending systems discussed by Voogd (this volume). Thus, this area has been investigated for its roles in the control of both movement and posture. In recent years, neural activity in this area has been extensively studied relative to the achievement of a behavioural state termed motor preparation or motor set (Wise, 1989). Using an instructed delay period paradigm, in which the animal receives instruction for a limb movement but before a signal allowing the movement to be executed, it has been shown that some PM cells discharge in directionally specified ways (Wise et al., 1983). When the same visual stimulus indicates withholding the movement, then there is hardly any set-related activity in PM neurons. All of this suggests that a particular class of cells in the PMC may adjust the excitability of cortical and subcortical networks so as to provide a background or postural set for initiating the upcoming movement. It may be the case then, that the PMC provides a preparatory 'tuning' of the axial and proximal musculature for particular visuomotor tasks. In other words, it forms a site where perception and posture are coupled to enable the performance of a task-specific action.

It is unclear if and how the PMC participates in the visual control of distal limb movements, although this role has been emphatically stressed (Rizzolatti et al., 1988). In humans, blood flow in the PMC (as well as in the SMC) increases during the execution of arm movements carried out on the basis of verbal instructions (Roland et al., 1980). This finding, along with PMC ablation studies in monkeys (Halsband & Passingham, 1982), has led to the tentative conclusion that this area provides an abstract form of movement guidance related to properties of the target (Wise, 1985). The ablation studies are interesting in that they revealed that the monkeys retained the ability to carry out sequences of movements and control of the fingers appropriate to specific tasks. This implies that the PMC is not involved in specifying the temporal and dynamic parameters of goal-directed movements.

In conclusion, the PMC appears to have two important functions in motor control. Firstly, it has a class of neurons that prepare the animal for an intended movement, perhaps through a task-specific coupling between perception and posture. Secondly, it is not involved in the control of individual movements, but rather in the guidance of actions relative to the specific demands of the task. In turning to the SMC, we find a similar assignment of functions, but some differences as well.

The SMC receives inputs from the basal ganglia via the thalamus. Together these structures form part of a generalised basal ganglia-thalamocortical circuit which centres on the putamen (Alexander et al., 1986). While the putamen receives afferent fibres from all motor fields of the cortex, it only projects back through thalamus to the SMC. Thus, the SMC has a privileged position in this circuit, both receiving outputs from and sending inputs to the neostriatum of the basal ganglia. Other projections from the SMC are to area 4 and to the arcuate neurons of the PMC. In addition, it has direct, if sparse, projections to the outer dorsolateral portions of the ventral horns in the lower cervical segments of the spinal cord. This finding led Alexander et al. (1986) to speculate that not only area 4 but also the SMC may have monosynaptic connections with spinal motorneurons innervating the hand muscles. The functional significance of this 'secondary' direct corticospinal pathway is presently unknown. In discussing the functions of the SMC, we should not forget that it forms, together with the basal ganglia, a very complex circuit which is becoming increasingly implicated in a wide range of skilled movements (Marsden, 1980).

Anatomically the SMC is situated upstream of area 4 and it is therefore possible that it influences the output of the primary motor cortex to cervical spinal motorneurons. One way this might be achieved is that the SMC acts to create a preparatory setting of responsiveness of the motor cortex to sensory inputs (Tanji et al., 1980). This suggestion implies that the SMC has a role in the initiation of movements in the distal segments. Support is provided from the observation that when human subjects are asked to rehearse, but not carry out, a sequence of finger movements there is an increase of cerebral blood flow in the SMC but not in area 4 (Roland et al., 1980). The same was found when subjects had to imagine opening and closing the hand (Fox et al., 1987). Further support stems from experiments where monkeys could select an action in advance: Under this condition many SMC cells discharged in the couple of seconds prior to initiation (Okano & Tanji, 1987). Thus, there may be a division of labour between the PMC and SMC, with the former providing a motor set for the axial and proximal musculature and the latter for the distal muscles of the hand.

A major difference between the SMC and the PMC was proposed on the basis of a synthesis of the findings from a number of animal experiments (Goldberg, 1985). It was hypothesised that the SMC plays an important role in internally generated movements and the PMC in externally cued movements. This hypothesis was only partially confirmed in a PET study on human subjects (Dieber et al., 1991). It was found that the SMC has significantly more cerebral blood flow in the tasks requiring the internal generation of movement than in those directed by an external cue (a tone indicating which one of four directions a joystick had to be moved). The PMC, however, was not more active during the externally cued tasks. These findings led to the conclusion that neural activity in the SMC is more dependent than that in the PMC on how the animal is prepared to respond to a particular task. This dependency may arise because of the involvement of the SMC via area 4 in the control of the distal segments rather than in specifying the direction of a movement.

4 Conclusions so far

The cortical structures discussed in this and Voogd's contribution suggest the following scenario for the cortical control of complex movements.

1. The premotor and supplementary cortices engender a task specific motor set.
2. The motor signal for initiating the movement is produced in the supplementary motor cortex.
3. The signal is sent to the premotor cortex where it is integrated with perceptual information.
4. The integrated information is received in the primary motor cortex which then activates the necessary muscles and controls the dynamic parameters of the movement.
5. The posterior parietal cortex controls the kinematic parameters through coupling the perceptual information generated by the moving limb with the kinetic parameters driving the musculoskeletal system.

The control of purposeful action is by no means the sole domain of the cortical motor fields. To illustrate this point we consider next the integrative action of the spinal cord given that Voogd has already dealt with other subcortical structures concerned with motor control. In doing so we turn to the role of the spinal cord in coordination of complex movements.

5 Spinal cord

Sherrington (1906) was at pains to stress the exquisite integrative capacity of the spinal cord. Following on from Pflüger and Vulpian, he stressed that the spinal cord was, in and of itself, a centre for coordinating movement. He drew this conclusion from his studies on the spinal dog. Despite the cord being separated from the rest of the brain, a mild cutaneous stimulus to the back resulted in alternating and rhythmical scratching movements with the appropriate hindlimb which was quite accurately directed towards the source of irritation. Graham-Brown (1911) attempted to find a synaptic explanation for such behaviour, an endeavour which led him, however, to propose that the spinal cord was capable of generating intrinsic rhythmical activity. This proposal foreshadowed the notion of central pattern generators and a shift away from the spinal cord as an organ of sensorimotor integration. Criticisms of this notion (reviewed in Post, 1992) have resulted in a reinstatement of sensory feedback in modulating the motor output of the spinal cord. Of interest here, is the claim that the spinal cord constitutes an important site for the neural integration of such actions as reaching.

At the cervical level there is a class of interneurons situated above the proximal motor nuclei. Given their location and hypothesised function, they have been termed the C3-C4 propriospinal neurons (Lundberg, 1979). A number of descending tracts, including both the corticospinal and the rubrospinal, make monosynaptic connections with these interneurons. In turn, they send axons to various pools of proximal motorneurons. When their output to these neurons is sectioned or their corticospinal input is removed, grasping is unaffected but reaching becomes abnormal in appearance (Alstermark et al., 1981). When peripheral input to the propriospinal neurons is interrupted overreaching the target is observed (Alstermark et al., 1986), thus implicating them in movement termination. Finally, it has been shown that the C3-C4 propriospinal system is active during reaching but not during locomotion - at least not on an even surface (Alstermark & Kümmel, 1986).

The anatomical, behavioural and physiological findings relative to this system have led to the following speculations (Georgopoulos & Grillner, 1989). Firstly, it plays an important role in the neural integration of goal-directed arm movements at the spinal level. Secondly, the initiation of an appropriately-directed reach is achieved by the activation of populations of neurons in the motor cortex, which then engages the spinal reaching circuits in the C3-C4 system. Thirdly, together with the cerebellum, which is indirectly connected with the propriospinal neurons via the lateral reticular nucleus (Alstermark et al., 1981), these circuits ensure coordinated reaching movements. Such speculations offer some solutions to Sherrington's quest as to how the cortex and the spinal cord cooperate in the control and coordination of action.

6 Concluding remarks

The traditional view of the primary motor cortex is that it is the motor area of the cortex rather than just the cortical area which is connected to the spinal cord (Diamond, 1979). In this contribution we have endeavoured to show that other cortical areas have important roles in motor control. These roles include the generation of motor sets, the initiation of complex movements, and the coupling of perceptual information with motor output in the regulation of purposeful actions. Thus, a clear picture is emerging of how the various areas of the cortex are functionally implicated in the control of movement and posture. It is also becoming clear that the spinal cord is not just a source of neural oscillators for rhythmical movements such as locomotion, but that it also plays a significant role in the neural integration of such discrete actions as reaching. Traditional neurophysiology from Sherrington onwards has attempted to understand how the brain performs these and other actions in terms of specific anatomical connections between neurons. In recent years, this view has given ground to one in which actions emerge from dynamical regulatory processes involving cooperative activity between many co-active neurons widely distributed throughout the nervous system (Szentágothai & Erdi, 1989). However, the more we discover about cortico-cortical and cortico-subcortical connections and their related functions, the more we will understand how these processes are channelled or constrained to produce goal-directed actions. The challenge for the future then, is to elucidate theories of neural dynamics that can assimilate the sorts of findings in functional neuroanatomy and behavioural neurophysiology as discussed in this and Voogd's contribution. Such theories will constitute a new agenda for studies on the integrative action of the nervous system.

Acknowledgements. We are grateful to Knoek van Soest for his comments on this paper and to Frank Zaal for his invaluable editorial assistance.

References

Alexander, G.E., DeLong, M.R., & Strick, P.L. (1986). Parallel organization of functionally segregated circuits linking basal ganglia and cortex. *Annual Review of Neuroscience*, **9**, 357-381.

Alstermark, B., & Kümmel, H. (1986). Transneural labelling of neurons projecting to forelimb motoneurones in cats performing different movements. *Brain Research*, **376**, 387-391.

Alstermark, B., Górska, T., Johannisson, T., & Lundberg, A. (1986). Hypermetria in forelimb target-reaching after interruption of the inhibitory pathway from forelimb afferents to C_3-C_4 propriospinal neurones. *Neuroscience Letters*, **3**, 457-461.

Alstermark, B., Lundberg, A., Norsell, U., & Sybirska, E. (1981). Integration in descending motor pathways controlling the forelimb in the cat: 9. Differential behavioural defects after spinal cord lesions interrupting defined pathways from higher centres to motoneurones. *Experimental Brain Research*, **42**, 299-318.

Crammond, D.J., & Kalaska, J.F. (1989). Neuronal activity in primate parietal cortex area 5 varies with intended movement direction during an instruction-delay period. *Experimental Brain Research*, **76**, 458-462.

DeFelipe, J., & Jones, E.G. (1988). *Cajal on the Cerebral Cortex: An Annotated Translation of the Complete Writings* . Oxford: Oxford University Press

Diamond, I.T. (1979). The subdivision of neocortex: a proposal to revise the traditional view of sensory, motor and association areas. *Progress in Psychobiology and Physiological Psychology*, **8**, 1-43.

Dieber, M.P., Passingham, R.E., Colebatch, J.G., Friston, K.J., Nixon, P.D., & Frackowiak, R.S.J. (1991). Cortical areas and the selection of movement: a study with positron emission tomography. *Experimental Brain Research*, **84**, 393-402.

Fel'dman, A.G. (1966). Functional tuning of the nervous system with control of movements or maintenance of a steady posture. III. Mechanographic analysis of execution by man of the simplest motor tasks. *Biophysics*, **11**, 766-775.

Fel'dman, A.G. (1986). Once more on the equilibrium-point hypothesis. *Journal of Motor Behavior*, **18**, 17-54.

Fox, P.T., Pardo, J.V., Petersen, S.E., & Raichle, M.E. (1987). Supplementary motor and premotor responses to actual and imagined hand movements with Positron Emission Tomography. *Society for the Neurosciences Abstracts*, **398**, 10.

Georgopoulos, A.P., & Grillner, S. (1989). Visuomotor coordination in reaching and locomotion. *Science*, **245**, 1209-1210.

Georgopoulos, A.P., Kalaska, J.F., Camaniti, R., & Massey, J.T. (1982). On the relations between the direction of two-dimensional arm movements and cell discharges in primate motor cortex. *Journal of Neuroscience*, **2**, 1527-1537.

Goldberg, G. (1985). Supplementary motor area structure and function: review and hypotheses. *Behavioral & Brain Sciences*, **8**, 567-588.

Graham-Brown, T. (1911). The intrinsic factors in the act of progression in the mammal. *Proceedings of the Royal Society, B*, **84**, 308-319.

Halsband, U., & Passingham, R. (1982). The role of premotor and parietal cortex in the direction of action. *Brain Reserch*, **240**, 368-372.

Jones, E.G., & Powell, T.P.S. (1970). An anatomical study of converging sensory pathways within the cerebral cortex of the monkey. *Brain*, **93**, 793-820.

Kalaska, J.F., Camaniti, R., & Georgopoulos, A.P. (1983). Cortical mechanisms related to the direction of two-dimensional arm movements: relations in parietal area 5 and comparison with motor cortex. *Experimental Brain Research*, **51**, 247-260.

Kalaska, J.F., Cohen, D.A.D., Hyde, M.L., & Prud'homme, M. (1989). A comparison of movement direction-related versus load-direction activity in primate motor cortex using a two-dimensional reaching task. *Journal of Neuroscience*, **9**, 2080-2102.

Kalaska, J.F., Cohen, D.A.D., Prud'homme, M., & Hyde, M.L. (1990). Parietal area 5 neuronal activity encodes movement kinematics, not movement dynamics. *Experimental Brain Research*, **80**, 351-364.

Kuypers, H.G.J.M. (1962). Corticospinal connections: postnatal development in the rhesus monkey. *Science*, **138**, 678-680.

Kuypers, H.G.J.M. (1964). The descending pathways to the spinal cord, their anatomy and function. *Progress in Brain Research, Organization of the Spinal Cord*, **11**, 178-200.

Kuypers, H.G.J.M. (1982). A new look at the organization of the motor system. *Progress in Brain Research*, **57**, 381-403.

Lundberg, A. (1979). Integration in a propiospinal motor centre controlling the forelimb in the cat. In H. Asanuma & V.S. Wilson (Eds.), *Integration in the Nervous System* (pp. 47-69). Tokyo: Igaku-Shoin.

Lynch, J.C. (1980). The functional organization of posterior parietal association cortex. *Behavioral & Brain Sciences*, **3**, 485-534.

Marsden, C.D. (1980). The enigma of the basal ganglia and movement. *Trends in Neurosciences*, **3**, 284-287.

Mountcastle, V.B., Lynch, J.C., Georgopoulos, A.P., Sakata, H., & Acuna, C. (1975). Posterior parietal association cortex of the monkey: command function for operations within extrapersonal space. *Journal of Neurophysiology*, **38**, 871-908.

Okano, K., & Tanji, J. (1987). Neuronal activity in the primate motor fields of the agranular frontal cortex preceding visually triggered and self-paced movements. *Experimental Brain Research*, **66**, 155-166.

Pandya, D.N., & Kuypers, H.G.J.M. (1969). Cortico-corticospinal connections in the rhesus monkey. *Brain Research*, **13**, 13-36.

Post, A. (1992). A systems theoretical analysis of the CPG concept. Master's Thesis, Free University, Amsterdam.

Rizzolatti, G., Camarda, R., Fogassi, L., Gentilucci, M., Luppino, G., & Matelli, M. (1988). Functional organization of inferior area 6 in the macaque monkey: II Area F5 and the control of distal movements. *Experimental Brain Research*, **71**, 491-507.

Roland, P.E., Larsen, B., Lassen, N.A., & Skinhj, E. (1980). Supplementary motor area and other cortical areas in organization of voluntary movements in man. *Journal of Neurophysiology*, **43**, 118-136.

Schell, G.R., & Strick, P.L. (1984). The origin of thalamic inputs to the arcuate premotor and supplementary motor areas. *Journal of Neuroscience*, **4**, 539-560.

Seal, J. (1989). Sensory and motor functions of the superior parietal cortex of the monkey as revealed by single-neuron recordings. *Brain, Behavior & Evolution*, **33**, 113-117.

Sherrington, C.S. (1906). *The Integrative Action of the Nervous System*. New York: Scribner.

Sherrington, C.S. (1933). *The Brain and its Mechanisms*. Cambridge: Cambridge University Press.

Strick, P.L., & Kim, C.C. (1978). Input to primate motor cortex from posterior parietal cortex (area 5). 1. Demonstration by retrograde transport. *Brain Research*, **157**, 325-330.

Szentágothai, J., & Erdi, P. (1989). Self-organization in the nervous system. *Journal of Social and Biological Structures*, **12**, 367-384.

Taira, M., Mine, S., Georgopoulos, A.P., Murata, A., & Sakata, H. (1990). Parietal cortex neurons of the monkey related to the visual guidance of hand movement. *Experimental Brain Research*, **83**, 29-36.

Tanji, J., Taniguchi, K., & Saga, T. (1980). Supplementary motor area: neuronal response to motor instructions. *Journal of Neurophysiology*, **43**, 60-68.

Wise, S.P. (1985). The premotor cortex: past, present and preparatory. *Annual Review of Neuroscience*, **8**, 1-19.

Wise, S.P. (1989). Frontal cortex and motor set. In M. Ito (Ed.), *Neural Programming, Taniguchi Symposia on the Brain Sciences, No. 12.* (pp. 25-38). Tokyo & Basel: Japanese Scientific Societies Press & Karger.

Wise, S.P., Weinrich, M., & Mauritz, K.H. (1983). Motor aspects of cue-related neuronal activity in premotor cortex of the rhesus monkey. *Brain Research*, **260**, 301-305.

Chapter 5: Neuronal network generating locomotor behaviour in lamprey: Circuitry, Transmitters, Membrane Properties, and Simulation[*]

S. Grillner[1] , P. Wallén[1], L. Brodin,[1] & A. Lansner[2]

1 Introduction

All patterns of behaviour are produced by interacting nerve cells. Although progress has been rapid on the level of the single nerve cell and its different types of ion channels, little or no knowledge is available on how the neural networks underlying different aspects of the vertebrate behavioural repertoire may function on a cellular level. The reason for this condition is that a detailed knowledge about the circuitry is required, for instance how different, relevant nerve cells interact, their properties, the types of synaptic interactions between interneurons, and so forth. Such detailed knowledge has been beyond reach with current techniques for these complex mammalian nervous systems, which have been studied in some detail, like those of rat and cat. Nevertheless, much valuable information has been gathered about these nervous systems concerning which parts of the brain are important for a given function and the general characteristics of individual control systems, pathways, transmitters, central networks, and types of sensory interaction. Between this general organisational level and the single cell level this spans a wide gap. It has been bridged in only a few instances, when vertebrate behaviour has been explained in terms of interacting nerve cells. Obviously the only way to remedy this condition is to find suitable vertebrate, experimental models with few neurons in the relevant circuits, and favourable conditions for analysis. Two lower vertebrate models, those of tadpole and lamprey, have proven advantageous (see Grillner et al., 1987, 1988a,b; Roberts et al., 1981, 1986; Cohen et al., 1988; Sillar & Roberts, 1988; Sigvardt, 1989). Much knowledge has been gained in a variety of invertebrate 'simple' systems (Burrows, 1989; Getting, 1988; Small et al., 1989; Selverston & Moulins, 1985), but this information is not transferable, when it comes to specific neural organisation, due to the major anatomical difference between vertebrate and different invertebrate nervous systems.

Our interest has been focused on one of the most basic patterns of vertebrate behaviour performed at will, that of locomotion, which is a very complex motor act. It is generated by a family of control systems providing propulsion, equilibrium/stability, steering, and adaptation to the environment. Although the locomotor movements of each species have their specific characteristics, and the types of locomotion differ markedly among walking, flying, and swimming, the control systems appear very similar throughout the vertebrate phylum (Grillner, 1981, 1985). The neural machinery for propulsion has been explored to some degree in all classes of vertebrates (see Jordan, 1986; Grillner & Wallén, 1985; Noga et al., 1988; Stein, 1978; Rossignol et al., 1988; Shik & Orlovsky, 1976; Orlovsky & Shik, 1976; Garcia-Rill & Skinner, 1986). This motor pattern is produced by networks of interneurons located in the spinal cord, which interact with sensory input. Neurons descending from the brainstem initiate and control the level of activity in the spinal cord networks, and thereby the speed of locomotion. Below we discuss the swimming behaviour of the lamprey in three parts, which we treat separately.

[*] Reproduced with permission, from the *Annual Review of Neuroscience* **vol. 14**: 169-199, 1991 by Annual Reviews Inc.

Authors' adress: 1) Nobel Institute for Neurophysiology, Karolinska Institute and 2) The Royal Institute of Technology, Stockholm, Sweden.

- The propulsive neural machinery.
- The equilibrium control system.
- The system for steering and adaptation to the environment.

The lamprey CNS is an experimentally amenable vertebrate model in which the brainstem and spinal cord can be maintained in vitro over a period of several days (Rovainen, 1979b). Although the lamprey's nervous system is simple, with fewer neurons than in other vertebrates, it has the ground plan of other vertebrates, including, for instance, telencephalon, diencephalon with basal ganglia, mesencephalon and rhombencephalon, the different cranial nerves, and the descending reticulospinal pathways. Since lampreys belong to the most 'primitive' vertebrate group (cyclostomes), comparative anatomists have shown a substantial interest in its nervous system (for review see Nieuwenhuys, 1977). The application of modern techniques in studies of the brainstem-spinal cord morphology (for review see Brodin & Grillner, 1990; Ronan, 1989; Northcutt, 1984) has allowed a more detailed analysis of different neuronal system with a transmitter identification. The knowledge of the different types of neurons studied with paired recordings may now be greater than in other vertebrates ((Rovainen, 1974a,b, 1979a; Buchanan, 1982; Buchanan & Grillner, 1987a,b; Ohta & Grillner, 1989; Viana di Prisco et al., 1989).

2 Actual swimming behaviour

The lamprey has no paired fins, and swims by propagating a mechanical, laterally directed wave along its long eel-like body (Figure 1A). The faster it swims the faster the wave is propagated down the body, pushing the rostral end forward through the water. The amplitude of the movement increases towards the tail. As it swims, it always keeps the body oriented with the back upwards, as in all fish. By bending the body somewhat further to the left or right, or up or down, it can steer the body with great precision towards different points in space.
Alternating bursts of activity in segmental motoneurons make the muscle fibres in each segment contract in an alternating fashion on the left and the right side. Each segment is coordinated with its caudal neighbour, such that the contraction cycle of the latter occurs somewhat later (1% of the duration of the swim cycle; cf. Figure 1B). Since the body has about 100 segments, the phase delays will accumulate along the body to a total phase lag of around 100% between the rostral and the most caudal segments. The result is that a wave is transmitted along the body. Since this phase delay is a given proportion of the swim cycle (1% per segment), the actual delay in seconds (Figure 1C,D) will vary substantially (40-fold). The swim cycle duration varies directly with the speed of swimming from around 4 s to 100 ms (i.e., 0.25 to 10 Hz; Wallén & Williams, 1984; Williams et al., 1989), and accordingly with the speed of caudal propagation of the travelling wave. The speed of swimming of all animals, of the same body length, will thus be similar at the same swim cycle frequency. If the body length differs, the distance in millimetres between segments will differ accordingly, and the speed of the travelling wave will exhibit a corresponding difference. Hence, the speed of swimming is proportional to the body length, everything else being equal (cf. Bainbrigde, 1963; Grillner & Kashin, 1976; Webb et al., 1984).

3 The neuronal circuitry underlying propulsion

The motor pattern underlying locomotion can be produced in the isolated brainstem spinal cord

The isolated brainstem spinal cord can be maintained in vitro (Rovainen, 1979b). By stimulation for reticulospinal neurons in the brainstem, the motor pattern underlying locomotion can be produced along the spinal cord with a maintained intersegmental

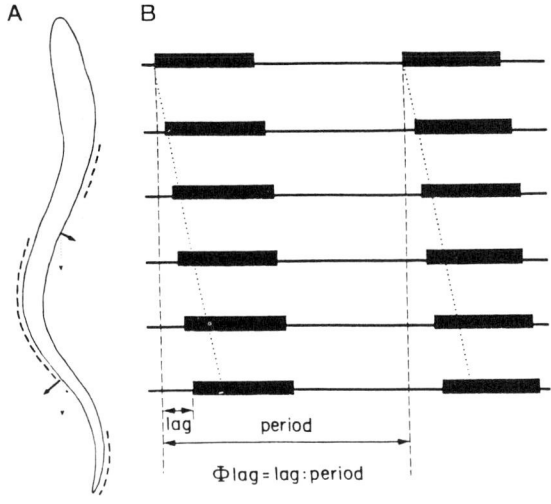

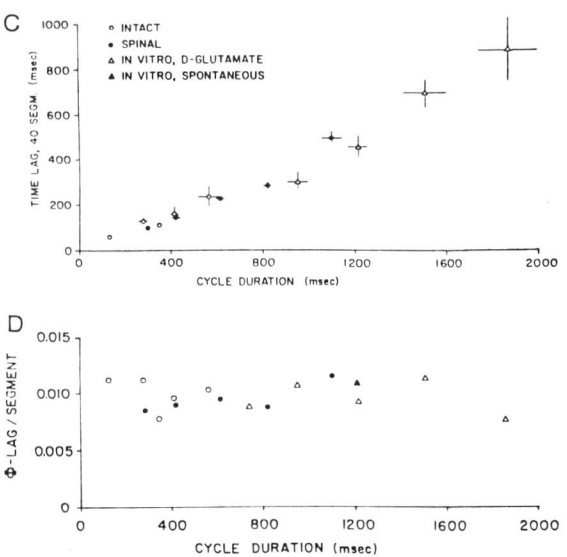

Figure 1. Characteristics of the actual swimming behaviour in lamprey. A. Body outline at one particular instance during swimming. Dotted lines denote regions of active muscle contraction. Arrows indicate forces being exerted against the water, with the caudally directed vector (dotted) propelling the animal forward. B. Schematic diagram indicating bursts of EMG activity in six segments on the rostral part of the body. There is a caudally directed electromyographical wave of contraction, resulting in a phase lag between consecutive segments. This intersegmental lag varies considerably in time with changing speed of locomotion (C; shown here for intact, spinal, and in vitro preparations), but constitutes a constant proportion of the cycle period (D), about 1% between adjacent segments, irrespective of the speed of swimming. (C, D: Modified from Wallén & Williams, 1984).

41

coordination (McClellan & Grillner, 1984). In the isolated spinal cord sensory input from tailfin afferents, spinal tract stimulation or application of excitatory amino acids in the bath (Figure 3A,B,C)
may elicit such a coordinated 'fictive' locomotor pattern (Cohen & Wallén, 1980; Grillner et al., 1981b; McClellan & Grillner, 1983; Poon, 1980). This motor pattern covers the same burst frequency range (0.25-10 Hz) and has an intersegmental coordination similar to that of normal swimming (Wallén & Williams, 1984; Brodin et al., 1985), although the variability of the pattern can be greater than in normal swimming. The spinal cord can be divided into several smaller parts down to around three segments, each of which can produce fictive locomotor activity with intersegmental coordination. If the pieces are divided further into parts of 1-2 segments, alternating activity can still be produced, although the variability of the burst pattern becomes much greater (Grillner et al., 1988b).

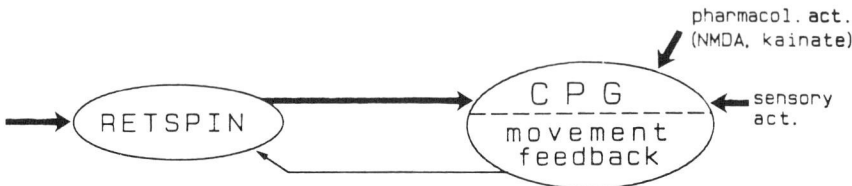

Figure 2. General vertebrate scheme depicting the different factors involved in the control of locomotion, with particular reference to the lamprey. Activation of reticulospinal neurons initiates and sets the level of activity in the spinal cord pattern-generating networks (CPG), which feed back efference copy signals to the reticulospinal neurons (which thereby become phasically modulated). Activity in the segmental networks can also be pharmacologically induced by activation of NMDA and kainate receptors. In addition, stimulation of sensory receptors (tail fin) can also activate the CPG networks.

    The circuitry underlying locomotion is comprised of three main components (Figure 2):
1. A spinal interneural network organised to provide segmental burst activity along the spinal cord with an intersegmental constant-phase coupling (Figure 3E).
2. A sensory control from stretch receptor neurons activated in each swim cycle, which can entrain the segmental locomotor network.
3. A descending reticulospinal system, which initiates spinal locomotor activity. It activates postsynaptic NMDA and kainate/AMPA receptors on all types of spinal interneurons in the network. During ongoing locomotion the reticulospinal neurons also receive spinoreticular locomotor-related input, which phasically modulates their burst activity.

Below we first deal with the segmental network, then its sensory control, and finally its supraspinal control.

The Segmental Circuitry and Its Mode of Operation

MOTONEURONS Motoneurons serve as pure output elements. Their actual output (pattern of action potentials) is a function of their membrane properties and their synaptic input. In each swim cycle they receive excitation via NMDA and kainate/AMPA receptors from spinal excitatory interneurons (Dale, 1986; Buchanan et al., 1989) during the excitatory half of the swim cycle, and glycinergic inhibition during the following half of the cycle (Russell & Wallén, 1983; see MN in Figure 3E). There is no evidence of synaptic interaction between the motoneurons themselves or from motoneurons to interneurons (Wallén & Lansner, 1984). Motoneurons supplying different parts of the myotome from dorsal to ventral have different dendritic trees, ramifying in partially not-overlapping areas, thereby suggesting a somewhat different origin of their synaptic input (Wallén et al., 1985).

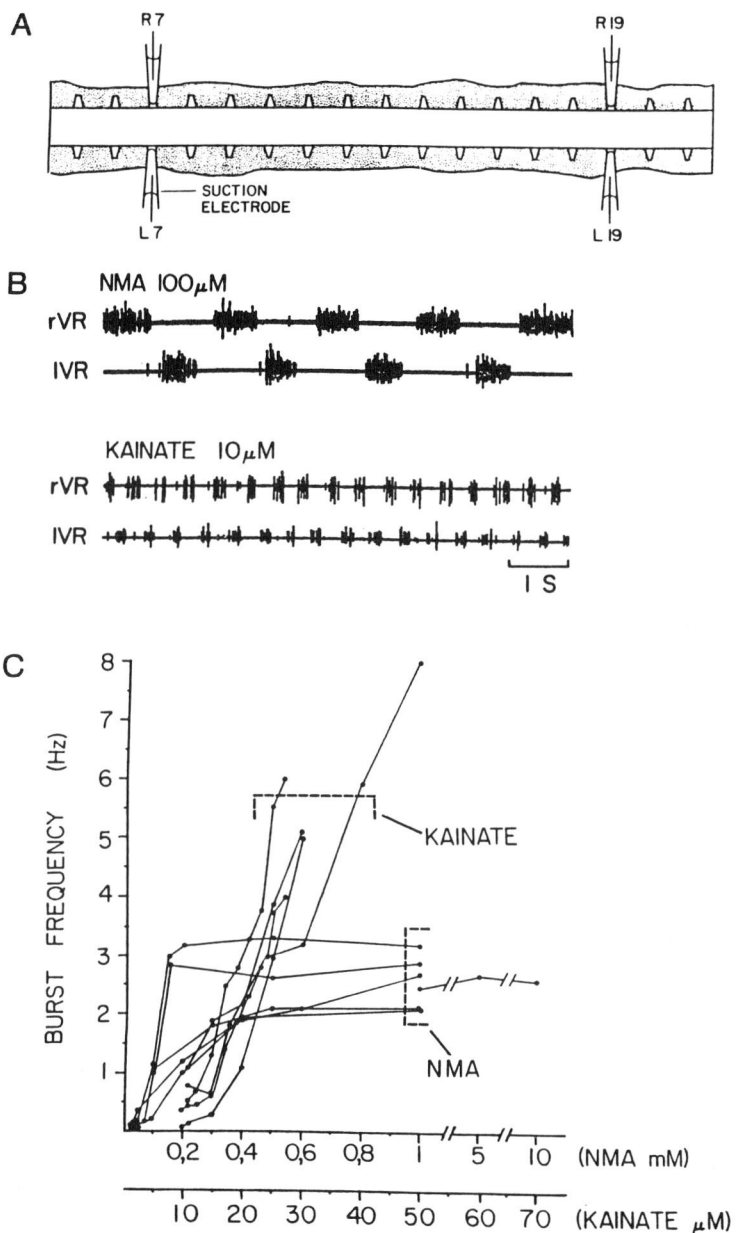

Figure 3. Initiation of fictive locomotion in the in vitro preparation of the lamprey CNS. A. Experimental set-up for recording of ventral root activity in the isolated spinal cord. B. Alternating bursts of activity recorded in two opposite ventral roots. Fictive locomotion initiated by bath application of N-methyl-D, L-aspartate (NMA) or kainate. C. Dose-response relationship of burst activity evoked by NMA and kainate, respectively. Increasing doses of NMA leads to bursting up to 2-3 Hz, where a plateau is reached, whereas increasing concentrations of kainate leads to a rapid rhythm up to 8-10 Hz.

43

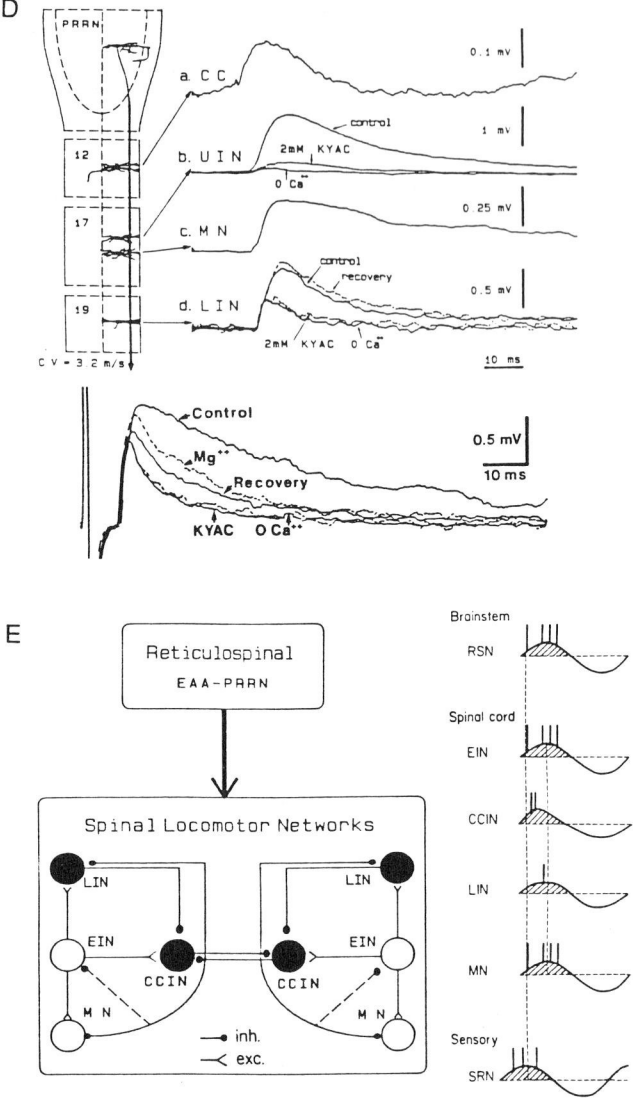

Figure 3 (continued) D. Activation of reticulospinal neurons evokes monosynaptic EPSPs in spinal neurons. Upper set of traces shows EPSPs evoked in several different spinal neurons upon intracellular stimulation of the same neuron in the posterior rhombencephalic reticular nucleus (PPRN: paired intracellular recordings). EPSPs are EAA-mediated since they could be blocked by the antagonist kynurenic acid (KYAC). In zero Ca^{2+} solution an electrical component could be seen (LIN). The reticulospinal EPSP could also be shown to have an NMDA component, which was blocked by Mg^{2+} ions (lower traces). E. The segmental interneuronal circuitry responsible for producing alternating burst activity. Abbreviations: EIN, excitatory interneuron; CCIN, contralaterally, caudally projecting interneuron; LIN, lateral interneuron; MN, motoneuron. The diagram to the right depicts the patterns of rhythmic activity in the network interneurons and in brainstem reticulospinal neurons (RSN), as well as in sensory stretch receptor neurons (SRN). See text for further details. (D: Modified from Ohta & Grillner, 1989).

EXCITATORY AND INHIBITORY PREMOTOR INTERNEURONS - SYNAPTIC INTERACTION IN THE PATTERN GENERATOR A network of premotor interneurons established experimentally with paired recordings is shown in Figure 3E. An excitatory type of premotor interneurons (EIN), acting through excitatory amino acid receptors, monosynaptically excites motoneurons and two types of inhibitory interneurons on the ipsilateral side (Buchanan & Grillner, 1987a; Buchanan et al., 1989). One commissural inhibitory interneuron inhibits contralateral motoneurons, EINs, as well as the inhibitory interneurons (Figure 3E). This interneuron has been referred to as the CC-interneuron (CCIN), since it has a long (> 20 segments), crossed, and caudally directed axon (Buchanan et al., 1982). There are also smaller crossed interneurons with a segmental distribution, which may provide additional reciprocal inhibition between the two sides (Ohta, Dubic & Grillner, unpublished; Wallén, unpublished; cf. Buchanan & Grillner, 1987b). The third type is the lateral inhibitory interneuron (LIN in Figure 3E, Rovainen, 1974a), which is fairly large with a long ipsilateral axon. These neurons are prominent in the rostral part of the spinal cord, but less conspicuous in the caudal half, presumably due to a smaller size. They inhibit caudally located ipsilateral CC interneurons and occasionally motoneurons (Rovainen, 1974a; Buchanan, 1982).

The pattern of activity during locomotion in these different types of neurons is shown to the right in Figure 3E. The neurons on the left and right side of the network have a reciprocal pattern: when one side is active the other is inhibited and vice versa. The EINs and motoneurons have a similar pattern of depolarisation followed by inhibition (Buchanan & Grillner, 1987a). The depolarising phase of the CCIN is terminated earlier (Kahn, 1982), due to the activation of the LIN, which has a higher threshold for eliciting an action potential (Buchanan & Cohen, 1982). This network of neurons can account for the rhythm generation when due consideration is given to membrane properties and types of synaptic interaction, as is shown below.

SIGNIFICANCE OF DIFFERENT TYPES OF SYNAPTIC INTERACTION - KAINATE/AMPA AND NMDA IN THE LOCOMOTOR NETWORK. Although the segmental burst-generating network normally receives a phasic excitatory drive from the brainstem and from segmental stretch receptor neurons in each swim cycle (see below), coordinated fictive locomotor activity may still be produced by a few isolated spinal cord segments, provided that the background excitation is at a sufficiently high level.

By applying the different agonists of NMDA and kainate/AMPA receptors in the perfusing fluid that passed over the isolated spinal cord, the fictive motor pattern can be elicited (Figure 3B; Grillner et al., 1981a; Brodin et al., 1985; Alford & Grillner, 1990). That a crude stimulus such as a superfusion of an excitatory amino acid over the spinal cord is sufficient to initiate a complex motor pattern may indeed seem surprising. The explanation is most likely simple; The spinal network is normally turned on by excitation of all neuronal elements (see below), and an application of excitatory amino acids in the bath will similarly activate all neurons with EAA receptors and thus all neurons of the segmental locomotor network. The actual motor pattern itself is produced by the inhibitory and excitatory interaction of the spinal interneurons discussed above. Activation of kianate (Brodin et al., 1985) or AMPA receptors (Alford & Grillner, 1990) after a blockade of NMDA receptors elicits locomotor activity in a range form around 1 to 8 Hz, which is in the upper part of the physiological frequency range (Figure 3C). In this case, kainate or AMPA produces the background excitation, and the excitatory interneurons (EINs) can in addition produce fast kainate/AMPA receptor mediated EPSPs, thereby contributing to the synaptic interaction within the network. If locomotion is instead initiated by sensory input or from the brainstem, a blockade of NMDA receptors will leave the faster locomotor activity undisturbed but block slow swimming.

An activation of NMDA receptors elicits fictive locomotor activity in a low-frequency range from 0.1 to around 2-3 Hz (Grillner et al., 1981b). Even if all fast kainate/AMPA transmission is blocked by the selective antagonists DNQX or CNQX, NMDA receptor activation can still be produce locomotor activity at the same burst rate as before the blockade (Alford & Grillner, 1990). In this case, the fast kainate/AMPA receptor-mediated synaptic interaction thus appears to have a limited role in pattern generation, which under these conditions depends on the inhibitory synaptic interaction and phasic NMDA receptor-induced pacemaker-like properties (see below).

The different types of EAA receptors appear very similar in mammals (Mayer & Westbrook, 1987) and lamprey (Dale & Grillner, 1986; Alford & Grillner, 1990), even though the different species of lamprey separated from the main vertebrate evolutionary line around 450 billions years ago. The NMDA receptor and its ion channel thus exhibits voltage dependence in the presence of physiological level of Mg^{2+} ions (Nowak et al., 1984; Grillner & Wallén, 1987; Moore et al., 1987; Hill et al., 1989). Moreover, a low level of glycine potentiates the activity of the NMDA receptor due to activation of a particular site on the ionophore (Ascher & Nowak, 1988), and a blockade at this site causes a marked depression of locomotion in lamprey, thus indicating that the glycine site is of behavioural significance (Grillner et al., 1990). The unique properties of NMDA receptors appear to play an important role in maintaining a slow steady rate of swimming (Brodin & Grillner, 1985, 1986). Apparently the voltage dependence of the NMDA receptor is reduced or abolished by washing out the Mg^{2+} ions, the slow regular swimming is depressed by the fast swimming is not affected, because NMDA receptors in the presence of Mg^{2+} ions can elicit plateau-like membrane potential shifts.

An NMDA receptor activation in a cell in which the action potential has been blocked by tetrodotoxin can give rise to pacemaker-like membrane-potential oscillations (Sigvardt & Grillner, 1981; Sigvardt et al., 1985; Grillner & Wallén, 1987). These oscillations (Figure 4A) are produced by an interaction between NMDA channels and other ion channels in the cell. A rapid depolarisation occurs as the voltage-dependent NMDA channels open up (2 in Figure 4A). The membrane potential reaches a plateau as the voltage-dependent K^+ channels are turned on by the depolarisation (3 in Figure 4A). The K^+ channels counteract the depolarising effect of the NMDA channels. Subsequently, another factor also comes into play: Ca^{2+} ions pass into the cell through the open NMDA channels, thus causing a progressive accumulation of Ca^{2+} and activation of Ca^{2+} dependent K^+ channels, which in turn will gradually pull the membrane potential in a repolarising direction (Figure 4A). At a certain level of membrane potential, the voltage dependence will cause the NMDA channels to close (4 in Figure 4A), a result that by itself will lead to a rapid repolarisation and subsequently a gradual recovery. These plateau-like depolarisations can be strongly influenced by other types of input and can be both shortened and prolonged and may thus become integrated into a network of interacting cells. These plateau-like potentials are of particular importance in relation to the generation of slow regular burst activity (see below).

NMDA receptors are thus activated in conventional synaptic transmission. This characteristic has interesting functional consequences due to the NMDA receptors' voltage dependence. At the resting membrane potential level, very small effects will occur, but at near-threshold for the action potential, the voltage-dependent NMDA channels will open and potentiate the effect of other inputs.

MODE OF OPERATION OF THE SEGMENTAL NETWORK To explore whether the segmental network (Figure 3E) of premotor interneurons alone without a sensory or supraspinal modulation can account for the pattern generation, mathematical modelling and computer simulations have been performed. The following experimentally supported suppositions have been made:

1. The network can be turned on by simply increasing the excitability in all neurons, such that they tend to discharge action potentials when not actively inhibited. Such a type of activation can be elicited by stimulation of the reticulospinal neurons activated from

locomotor areas, by sensory input (tailfin, trigeminal nerves), or by adding excitatory amino acids to the bath.

2. The pattern of inhibitory and excitatory synaptic interaction within the spinal network, combined with the dynamic cellular properties of network interneurons, forms the substrate for the generation of the alternating motor pattern produced.

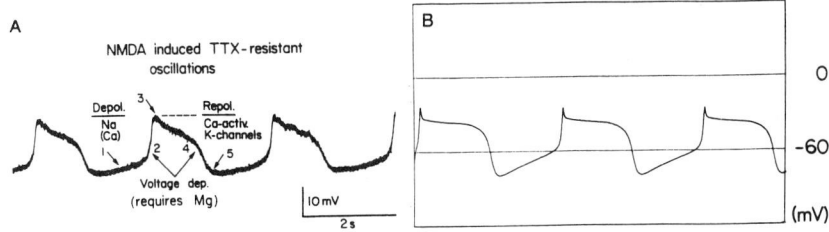

Figure 4. NMDA receptor-induced membrane potential oscillations in lamprey neurons. A. Experimental recordings in the presence of tetrodotoxin (TTX). Opening of the voltage-sensitive NMDA channels results in a depolarising current, mainly carried by Na+ ions. Ca2+ ions also entering through the NMDA channels will activate Ca2+-dependent K+ channels, which leads to repolarisation and a plateau. The subsequent closing of the NMDA channels will then cause a further repolarisation. With decreasing levels of intracellular Ca2+, the hyperpolarising current will decay and another depolarisation phase will start, B. Computer simulation of pacemaker-like NMDA induced membrane potential oscillations (see text).

The normal pattern of discharge in the neurons of the network is represented in Figure 3E. Let us consider the possible function of the network components in a general sense before we discuss the simulations. Thus, if the excitability is increased and one side of the network starts to become active, the neurons of the contralateral side will automatically become inhibited by the commissural CC-interneurons. If the background excitability remained the same, the asymmetric discharge could theoretically continue forever, and consequently no rhythmic burst activity would result. For alternating burst activity to occur, efficient burst-terminating factors must be in operation, which act to block the activity of the ipsilateral CC-interneuron and thereby disinhibit the neurons on the contralateral side, which will then become active due to their high background excitability. Several factors appear to contribute to the burst termination: (a) frequency adaptation due to summation of the afterhyperpolarisation; (b) the lateral interneuron (LIN), which becomes active in mid-cycle and directly inhibits the CC-interneurons; (c) the termination of the NMDA plateau in CCINs (and directly in EINs), which will obviously end the period of CCIN-activity. General considerations of this type provide a picture of possible mechanisms but provide no solid evaluation of the relative role of different mechanisms or whether the different cells can indeed interact to produce an appropriate pattern. Such an evaluation can only be obtained by computer simulation.

The network of Figure 3E has been modelled. Each neuron is simulated as a Rall-model with one soma and a three-compartment dendritic tree. Voltage dependent Na+, K+, and Ca2+ channels are simulated with Hodgkin-Huxley formalisms, and Ca2+-dependent K+ channels are also simulated. Inhibitory input synapses are represented by a conductance increase for Cl- and conventional excitation is represented by a conductance increase for Na+ and K+ (equilibrium potential of around 0 mV). In addition, voltage-dependent NMDA channels can be simulated (Grillner et al., 1989b) including the pacemaker-like oscillations (Figure 4B). Each type of neurons is given membrane properties and input resistance values similar to those observed experimentally. The shape of the action potential and the early and late phases of the afterhyperpolarisation are made to match those of the real cells. The synaptic potentials are similarly matched in terms of rise time and shape. Below we first discuss simulation of the segmental network, and subsequently discuss the sensory and supraspinal circuitry.

Simulations with three neurons of each type are shown in Figure 5 (that is, altogether 18 neurons; Grillner et al., 1988a) and with one neuron of each type in Figure 5A. Each type of interneuron is represented, their combined activity resembles that of the biological network (cf. Figure 3E). By modifying the background excitability in all neurons, the frequency of bursting changes from around 1 to 10 Hz, in a more or less linear fashion (Figure 5B), as is the case with kainate-induced swimming as described above (Figure 3C). In the discussion above we have predicted that the real network would use LIN inhibition on the CCIN as one mechanism for burst termination. If the LIN inhibition is set to 0 in the simulations, the burst rate is reduced but not abolished, and thus the model-LINs contribute to the burst terminations (not illustrated; Grillner et al., 1988a, 1989). On the other hand, the fact that bursting continues shows that at least one other mechanism would be spike-frequency adaptation through summation of the long-lasting afterhyperpolarisation that follows the action potentials, whereby the second interspike interval becomes longer than the first and so forth. If this were to be the case, one would expect that a reduction of the afterhyperpolarisation would lead to longer bursts, since the summation of afterhyperpolarisation would be less due to a smaller amplitude, everything else being equal. Figure 5D shows that a reduction of the afterhyperpolarisation in the model neurons indeed causes a prolongation of the burst cycle. Thus it appears likely that the afterhyperpolarisation summation can serve as a burst-terminating factor. That this may indeed be the case is also suggested by the cellular effect of 5-HT (see below), which acts through a reduction of the afterhyperpolarisation amplitude in network neurons and also causes a prolongation of the burst cycle duration (Harris-Warrick & Cohen, 1985; Van Dongen et al., 1986a; Wallén et al., 1989).

If NMDA receptors with voltage-dependent properties are introduced into the interacting model neurons, the network model acquires the ability to generate slow rate bursting (Figure 5A). If bath-applied NMDA receptor activation is simulated, the network covers a frequency range from around 0.3 to 1.5 Hz, at which point it reaches a plateau and does not increase further. The real network respond similarly to increasing levels of NMDA (Figure 3C; Grillner et al., 1981a). This is most likely due to the NMDA-depolarising plateaus, which are difficult to terminate in their first part but more labile in their second part. The actual NMDA-induced pacemaker-like "membrane potential oscillations" in single cells can indeed be simulated in detail with the suppositions given above (Figure 4B). The individual cells in the network (EIN, CCIN) can respond to these NMDA plateaus, and, in the network situation, these plateau properties manifest themselves in the ability to generate longer bursts and thus a slow and steady burst rate (Wallén & Grillner, 1987).

— In conclusion, computer simulations have shown that the segmental circuitry so far described can account for the rhythm generation in a single segment. Although these findings are significant, other aspects are added as we consider the intersegmental coordination and presynaptic inhibitory mechanisms as well as sensory and supraspinal control (see below).

Sensory control of locomotion - Effects on segmental circuitry

In practically all cases tested, sensor signals from the moving body, be it the hindlimb of a cat (Grillner & Rossignol, 1978; Duysens & Pearson, 1980) or the trunk of a fish (Grillner & Wallén, 1977, 1982), have been found to have a powerful effect on the central locomotor network. The function of this input is to adapt the motor output to the actual movements, which can meet with unpredicted perturbations. Indeed, an important effect of the sensory input is to affect the timing of the different phases of the locomotor cycle such as the transition from extension to flexion in the step cycle or from left-to right-side activity during swimming in fish. The sensory input contributes to the actual timing but is not the sole decisive factor. Recall that the central network itself reliably switches from one phase to the next even without sensory input. Normally, however, the central network activity is integrated with the sensory input into a complete pattern generator circuit.

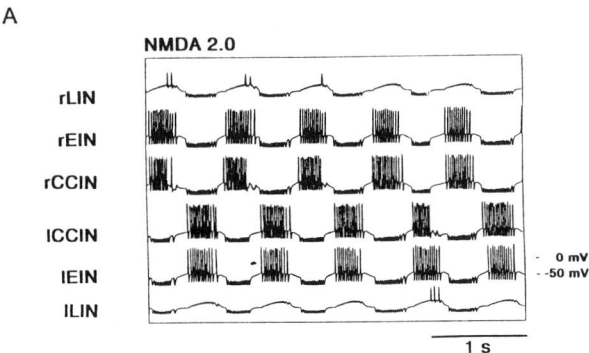

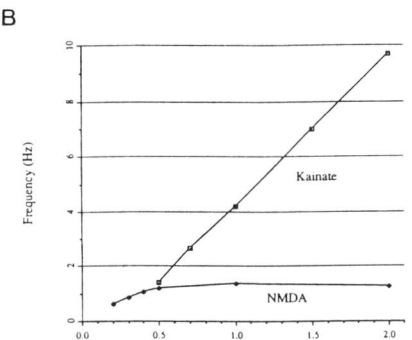

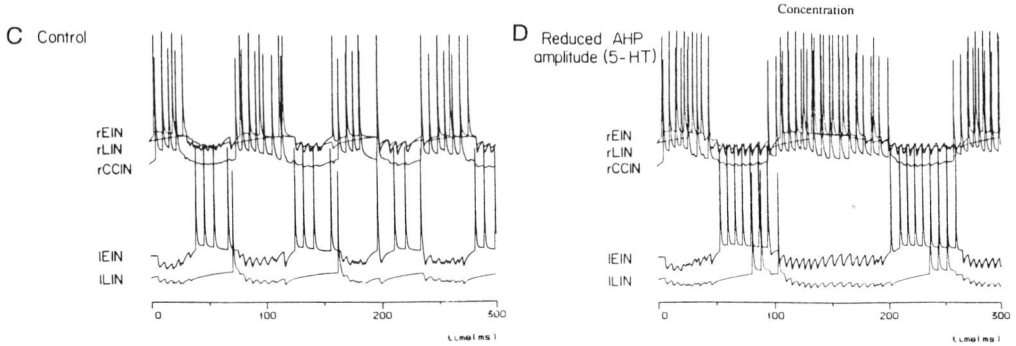

Figure 5. Computer simulation of the segmental network for locomotion in lamprey. A. Simulation of NMDA-receptor induced locomotor activity. Bath application of NMDA was simulated by a tonic opening of NMDA channels. The NMDA induced properties contribute to the burst termination (LIN activity not required). B. Dose-response relation between the level of kainate or NMDA receptor activation (arbitrary units) and the resulting burst frequency of the simulated network. The correspondence with experimental data is obvious (cf. Figure 3C). C, D. Simulation test of the role of the afterhyperpolarisation (AHP) as a burst-terminating factor. A reduced AHP amplitude leads to prolonged bursts, due to less summation of AGPs (D), which corresponds to the physiological effect of 5-HT.

49

In the case of the lamprey, intraspinal stretch receptor neurons (edge cells) sense the lateral bending movement that occurs during swimming, such that when one half of the myotome is contracting, the stretch receptor neurons on the contralateral side become extended and respond with a signal roughly proportional to the displacement (Grillner et al., 1981a,b, 1984; Grillner & Wallén, 1982). The stretch receptor neurons are located in the most lateral aspect of the flattened spinal cord, which obviously will be as stretched and unloaded as all other tissue when the body is bending. The stretch receptor neurons (Figure 6D) are of two kinds: (a) one with an ipsilateral axon, which provides monosynaptic EAA-mediated excitation to ipsilateral interneurons and motoneurons in the locomotor network, and (b) one with a crossed axon which is glycinergic-inhibitory to contralateral CCINs and other cells (Viana di Prisco et al., 1989). Thus, as the left side of myotome is contracting, the right side stretch receptor neurons will become progressively more extended and thereby activated. Their synaptic effects will be to inhibit the contralateral neurons (Figure 6D), which generate the ongoing contraction, and conversely to excite the ipsilateral network neurons, which should generate the next ipsilateral burst. Hence, it is apparent that the sensory input is organised to coordinate the central network activity with the actual ongoing movements. Thus, if the contraction on one side is very efficient, the stretch receptor neurons on the contralateral will become activated earlier than in other substances and consequently will promote an earlier transition of burst activity to the previously inactive side. Actually, to analyse the effect of the sensory input during ongoing fictive locomotion, the spinal cord preparation can be bent back and forth, while it rests the notochord, thus mimicking the movements occurring during locomotion (Grillner et al., 1981b). The superimposed movements will reliably entrain the central network to the extent that the ventral root burst activity follows (McClellan & Sigvardt, 1988) the superimposed movement over a range of approximately ± 50% of the resting cycle duration of the network

Figure 6A-C shows simulations of a spinal locomotor network (Figure 3E) to which the stretch receptor neurons have been added. (Figure 6D). Panel 6B shows the burst activity while the stretch receptors are silent. Swimming movements have been simulated by providing a sinusoidal current injected into the stretch receptor neurons on the left and right sides (Figure 6A, tract rEC, lEC), which are mase to be 180° out of phase with each other. This sinusoidal input corresponds approximately to that occurring during normal movements. The superimposed sinusoidal stretch receptor activity will effectively entrain the model network at rates both above and below the resting rates (Figure 6A-C). In addition, the relation between 'movement' and central network burst activity changes with frequency in similar fashion in both the real and the simulated entrainment. At a slow entrainment frequency, the bursts will occur early in the movement cycle in relation to the resting burst activity, but as the 'movement' frequency increases, the bursts will appear progressively later in the cycle, as expected from a system of coupled oscillators.

The simulation of the sensory input, which has been established experimentally, shows that the connectivity of the stretch receptor neurons with the neurons of the central network is sufficient to account for the sensory entrainment and the segmental pattern generation (Lansner et al., 1990).

Brainstem initiation of locomotion

Stimulation of reticulospinal neurons (Rovainen, 1974b; Wickelgren, 1977; Currie & Carlsen, 1987) in the lamprey brainstem can elicit fictive locomotion by activation of the spinal cord network (McClellan & Grillner, 1984; McClellan, 1988). Reticulospinal neurons from the posterior, middle, and anterior rhombencephalic reticular nuclei and the mesencephalic reticular nucleus include fast (Müller cells) and more slowly conducting neurons that operate through an activation of excitatory amino acid receptors. They provide monosynaptic

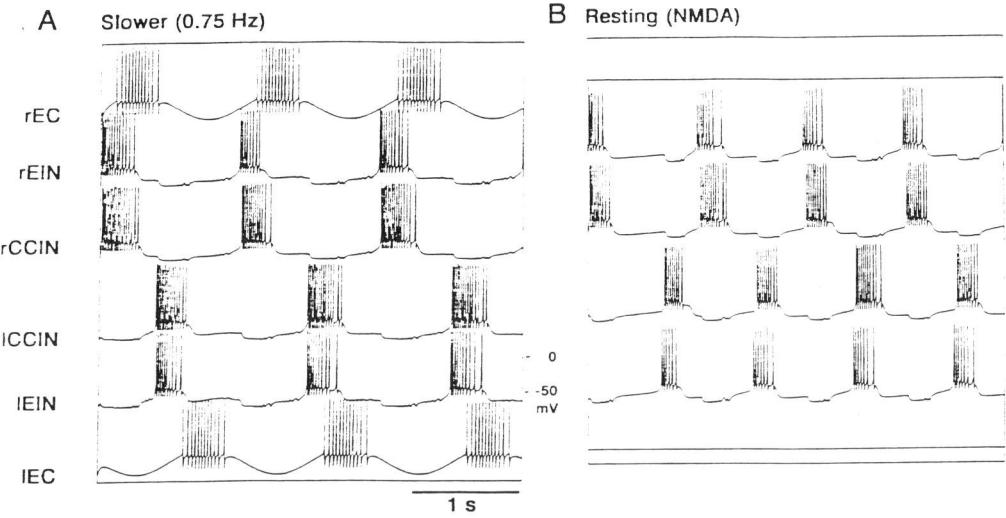

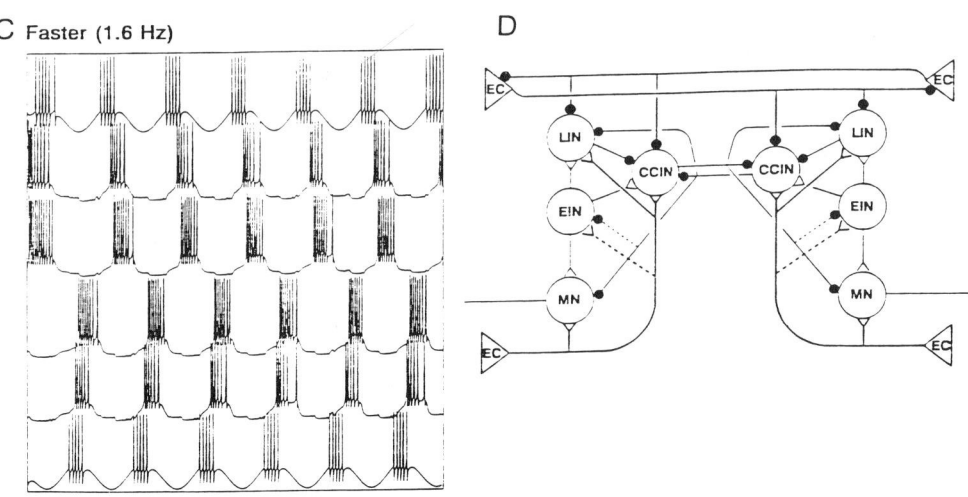

Figure 6. Simulation of the sensory and supraspinal control of segmental locomotor network. A-C: Simulated sensory entrainment of the network rhythm. A sinusoidal current was applied to model edge cells (EC) with a frequency slower (A) or faster (C) than the 'rest' rate (B). D: Diagram of the movement-sensitive feedback control of the segmental network, with ipsilateral projecting, excitatory edge cells and contralaterally projecting, inhibitory edge cells with connections as established experimentally.

E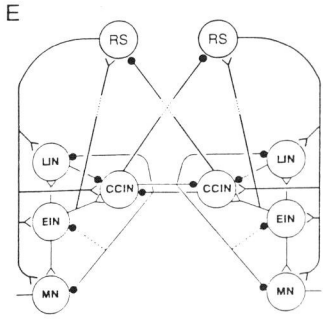

F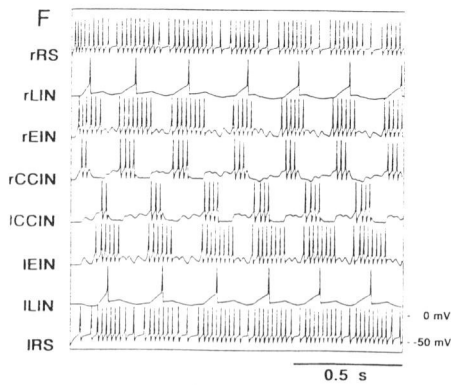

Figure 6 (continued) E, F. Simulation of reticulospinal initiation of locomotor activity in the segmental network. E. Two model reticulospinal neurons (RS) excite all neurons on either side of the network via EAA transmission (cf. Figure 3D). F. Resulting burst activity in the network, with kainate transmission only. Including NMDA transmission leads to slower burst activity.

excitation (Rovainen, 1974a) by a combination of NMDA, kainate/AMPA (Figure 3D), and often electric transmission (Brodin et al., 1988b; Ohta & Grillner, 1989). In addition to the EAA projections there are descending systems, which exhibit immunoreactivity towards 5-HT, PYY, and CCK, the functions of which are not yet clear (Brodin et al., 1986, 1988a,b).

Stimulation of the posterior or the middle reticular rhombencephalic nuclei may elicit fictive locomotion through a combined activation of NMDA and kainate/AMPA receptors (Figure 3D), a finding that is important in relation to the above discussion of EAA initiation of activity in the spinal network. If reticulospinal cells of different conduction velocities are recorded intracellularly, they increase their activity markedly when locomotion is initiated spontaneously or by sensory or central stimuli. The activity continues during the period of locomotor activity as recorded in the spinal cord. After the initial activation, however, most reticulospinal cells become modulated in phase with the ipsilateral locomotor activity in the rostral segments (Kasicki et al., 1988). The modulation results from ascending information from the spinal locomotor circuits (efference copy type), which provide supplementary alternating phasic excitatory (EAA) and inhibitory (glycinergic) synaptic drive. The axons ascend in the lateral funiculi with a relay interneuron located in the alar plate, as in the case of the dorsal column relays (Dubuc & Grillner, 1989; Dubuc et al., 1988). The different neurons on the posterior and middle reticular nuclei are very closely coupled to each other in terms of the intracellular oscillations. Thus the reticospinal cells, in a sense, can be considered as a part of the oscillating pattern-generating circuitry as indicated in the simplified scheme of Figure 6E, since they receive feedback from the spinal cord. If an extrinsic input provides an extra excitation of the reticulospinal neurons, they will start a bout of locomotor activity and the feedback will then act to reinforce and perpetuate or at least prolong the activity.

Normal initiation of locomotion in the behaving animal is elicited by the reticulospinal route, and we have thus explored whether locomotion can be elicited in simulations of the brainstem-spinal cord network (Figure 6E). We have used several configurations. Just adding excitation to the simulated reticulospinal neurons to make them discharge is sufficient to 'excite' the spinal network to generate simulated locomotor activity, even without spino-reticular feedback. When such feedback is added, however, the reticulospinal activity becomes phasic, and the entire motor pattern becomes more stable (Figure 6F). Thus, we conclude that the brainstem-spinal cord circuitry can provide the cellular basis for initiation and maintenance of segmental locomotor activity.

Different transmitters or putative transmitters involved in the modulation of the locomotor network

The synaptic interactions in the central and sensory locomotor network that we have simulated above is entirely mediated by postsynaptic excitatory EAA-transmission and glycinergic inhibition. Other systems, with transmitters like 5-HT and GABA and putative transmitters like somatostatin, CCK, and PYY, are also implied.

EXCITATORY AMINO ACID TRANSMISSION Excitatory amino acid synaptic transmission (presumably L-glutamate and/or aspartate) (Watkins & Evans, 1981) has in all cases investigated been found to be responsible for eliciting conventional EPSPs (kainate/AMPA and NMDA) in the lamprey CNS and in the locomotor circuitry studied (see Brodin, 1989). This applies to sensory dorsal cells (Brodin et al., 1987; Christenson et al., 1988a,b), excitatory stretch receptor neurons (Viana di Prisco et al., 1989), excitatory interneurons (Dale & Grillner, 1986; Buchanan & Grillner, 1987a), reticulospinal neurons (Buchanan et al., 1987; Ohta & Grillner, 1989), and oscillating spino-bulbar neurons (Homma, 1981).

GLYCINE Conventional postsynaptic inhibition, from for example, CC interneurons, inhibitory stretch receptor neurons, and smaller inhibitory interneurons, is blocked by strychnine and thus is presumably glycinergic (Viana di Prisco et al., 1989; Buchanan, 1982). Conversely, glycine opens Cl-channels (Homma & Rovainen, 1978; Kahn, 1982). The reciprocal inhibition between the left and the right side is a critical part of the pattern generation, and it is glycinergic. After a blockade of glycine receptors, rhythmic locomotor activity is blocked (Grillner & Wallén, 1980). After a partial blockade, however, bursting may remain on each side, but the two bursts instead become synchronous (Cohen & Harris-Warwick, 1984). Such burst activity is dependent on NMDA receptor activation (Alford & Sigvardt, 1989; Alford & Williams, 1989).

GABA$_A$ AND GABA$_B$ RECEPTORS ARE INVOLVED IN PRESYNAPTIC MODULATION AND ARE OF IMPORTANCE FOR THE INTERSEGMENTAL COORDINATION The axons of excitatory and inhibitory premotor interneurons have both GABA$_A$ and GABA$_B$ receptors (Alford et al., 1990); moreover, these receptors are phasically activated by GABAergic neurons in the spinal cord (Brodin et al., 1990a). An activation of either type of GABA receptor leads to a depression of the synaptic transmission to the postsynaptic cell, and thus GABA$_A$ and GABA$_B$ receptors take part on a presynaptic modulation. The axons are subject to a phasic depolarisation during each ipsilateral locomotor burst and thus presumably to a presynaptic inhibition of the transmitter release.

If either GABA$_A$ or GABA$_B$ receptors are blocked, no clear effects can be observed on the motor pattern (Grillner & Wallén, 1980; Alford et al., 1989; Alford & Grillner, 1990). Only if both types of GABA receptors are blocked at the same time (bicuculline and phaclophen) does the burst pattern become more variable, provided that comparatively long pieces of spinal cord comprised of many segments are used. If, however, short pieces of spinal cord (e.g., five segments) are used, no or very small effects are observed. Consequently, these GABAergic effects appear to be important whenever there are high demands on a well-functioning intersegmental coordination. In small pieces, all segments have a practically synchronous activity. The physiological role may thus be related to a segmental presynaptic control of the synaptic input of the segment that occurs mainly during the excitatory half cycle. Many interneurons like the inhibitory CC interneurons have axons extending over more than 20 segments - and each axon has a constant conduction velocity. Consequently, the inhibition from a given CC interneuron will reach a neuron 20 segments further down the cord after a fixed conduction delay. As the animal changes from slow swimming with a cycle period of 2 s to fast swimming at 0.1 s, however, the intersegmental phase delay of 20% (20 segments) of the cycle duration will vary in real time from 0.4 down to 0.02 s. Thus, the inhibitory CCIN signal could arrive at approximately the appropriate phase of the same cycle at one speed of swimming but at an entirely inappropriate phase at another speed. The ability of the segment to be able to 'gate' the input would thus seem very important, such that signals arriving in an inappropriate part of the movement cycle would be gated away. Interneuronal axons are thus modulated more or less in phase with the burst activity on the same side. That the reticulospinal drive signals from the brainstem are not subject to presynaptic GABAergic modulation is interesting and seemingly appropriate (Alford et al., 1990).

The sensory dorsal cell somata are not subject to phasic locomotor-related depolarisation, but a phasic input at the axonal level, which has been found in many other systems, cannot be excluded (Dubuc et al., 1985; Cattaert et al., 1990). GABA$_B$ receptor activation will cause a depression of the synaptic transmission from dorsal cells to interneurons (Christenson & Grillner, 1987 and unpublished).

There are two types of interneurons with immunoreactivity (ir) towards GABA in the spinal cord (Brodin et al., 1990a). Small, dorsally located, bipolar interneurons form varicosities in close apposition with dorsal cell axons, and perhaps mediate primary afferent control. A second, larger type located in the gray matter probably provides the rich supply of GABAir fibres in the lateral cell columns, and these fibres contact presynaptically modulate axons (Christenson et al., unpublished). From the above findings we would predict that such GABAir neurons are phasically active during locomotion. In addition, there are ciliated cells around the central canal with GABAir combined with somatostatin-ir. They extend processes to the lateral margin around the dendrites of the stretch receptor neurons (Brodin et al., 1990a,b; Christenson et al., 1989). Both GABA and somatostatin cause an inhibition of the stretch receptor neurons, in the first case, via Cl$^-$ channels (GABA$_A$); in the second case, probably via a K$^+$ channel mechanism.

5-HT SYSTEMS MODULATE LOCOMOTOR ACTIVITY BY MODIFYING CA2+-DEPENDENT K+ CHANNELS AND THEREBY THE NEURONAL FREQUENCY REGULATION The lamprey has descending 5-HT projections from the anterior and posterior rhombencephalic nuclei, and an intraspinal group of 5-HT cells located ventral to the central canal with bilateral ramifications, which form very dense ventromedial 5-HT plexa (Homma, 1970; Baumgarten, 1972; Van Dongen et al., 1985a,b; Harris-Warrick et al., 1985; Brodin et al., 1986). Medial dendrites of motoneurons and premotor interneurons like CCIN, LIN, and probably EIN have rich ramifications within this area. 5-HT varicosities may be in close apposition to the dendrites in the light microscope, but in the electron microscope no synaptic specialisations to indicate point-to-point synapses are found (Christenson et al., 1990). In addition, there are a few small cells with 5-HTir in each dorsal root ganglion and also in the cranial nerve ganglia, which presumably are sensory (Van Dongen et al., 1985a; Viana di Prisco, 1989). With regard to the brainstem, there are profuse 5-HTir ramifications in both the alar and the basal plate and around cell bodies of reticulospinal neurons, particularly in the rostral reticular nuclei.

5-HT applied to spinal cord or reticulospinal neurons, associated with 5-HT varicosities, elicits a depression of the afterhyperpolarisation following the action potential, but in most cases elicits no direct or indirect excitatory or inhibitory effect on the resting membrane potential. Since the afterhyperpolarisation is the main factor that determines the frequency regulation (Gustafsson, 1974), an afterhyperpolarisation depression has striking effects on the discharge pattern. The net effects of 5-HT is to make the cell fire at a higher rate at the same excitatory synaptic drive (Van Dongen et al., 1986a; Wallén et al., 1989). The 5-HT effect in the afterhyperpolarisation is exerted on the Ca^{2+}-dependent K$^+$ channels (which underlie the afterhyperpolarisation) presumably directly, rather than via a second messenger (Wallén et al., 1989). Some cells may also show some degree of hyperpolarisation, which appears to be due to indirect effects on other cells.

If 5-HT is applied in the bath during ongoing locomotion, the bursts will become more intense and longer and the burst rate will increase (Grillner & Wilén unpublished; Harris-Warrick & Cohen, 1985). Endogenously released 5-HT (blockade of 5-HT uptake) will give rise to similar effects on the network (Christenson et al., 1989b). The effects on the locomotor network can be explained entirely by the 5-HT effects on afterhyperpolarisation. As shown in the simulations described above (Figure 5C,D), a reduction of the amplitude of the afterhyperpolarisation will, everything else being equal, cause less summation of the afterhyperpolarisations and thereby diminish the spike frequency adaptation, which is one of the burst-terminating factors considered above.

In addition to the 5-HT effects on the AHP, 5-HT causes a depression of the EAA synaptic transmission from reticulospinal Müller cells to postsynaptic spinal neurons. 5-HT most likely interferes, either directly or indirectly, with the presynaptic release of transmitter, but it appears not to affect the activation of EAA receptors on the postsynaptic site (Buchanan & Grillner, 1987a,b).

PEPTIDERGIC SYSTEMS Combined studies with immunohistochemistry and biochemistry have shown that different types of tachykinins (Van Dongen et al., 1986b), cholecystokinin (CCK; Brodin et al., 1988; Ohta et al., 1988), pancreatic polypeptides (PYY, NPY), and neurotensin are present in the lamprey nervous system (Brodin et al., 1990b; Bongianni et al., 1989). The amino-acid sequences of the peptides are not yet known. There is immunoreactivity towards a number of additional peptides like somatostatin and CGRP (Buchanan et al., 1987b; Andrews et al., 1988; Christenson et al., 1989a). Except for the hyperpolarising effect of somatostatin on edge cells, discussed above (Christenson et al., 1989a), the functional effects of the peptide is unclear.

Intersegmental coordination - Possible neuronal mechanisms

Intersegmental coordination is characterised by a constant phase lag of around 1% of the swim cycle duration between adjacent segments along the spinal cord (see above and Figure 1). Since the cycle length varies between 4 and 0.1 s, the actual time delay varies accordingly. Consequently, the phase lag cannot be explained by conduction delays for intersegmental axons, but rather by other mechanisms (Grillner, 1974; Grillner et al., 1986a, 1988b). The intersegmental coordination does not require sensory feedback (Grillner et al., 1976; Wallén & Williams, 1984), and must be due to propriospinal interaction along the cord.

If we make the simplifying assumption that each segment contains one discrete pattern-generating network of oscillator (Grillner, 1974; Grillner et al., 1986a, 1988a,b), we can consider a portion of the spinal cord, e.g., ten segments, as a series of ten coupled oscillators. If each oscillator were independent, but all had approximately the same excitability, the oscillators would be active at approximately the same swimcycle period (e.g., 1 s). If the oscillators were connected to each other with even very weak excitation or crossed inhibition, they would tend to become coordinated. All segments would most likely become approximately synchronous and thus without a phase lag. If the excitability varied along the chain of oscillators, however, such that oscillator A if isolated would tend to go at a period of 1 s but oscillator B at 1.1 s, then oscillator A could entrain oscillator B, if they were connected in a unidirectional way from A to B. Since the inherent rate of B was somewhat slower than A, oscillator B would lag the activity of A in much the same way as is illustrated in Figure 6C with the sensory entrainment (extrinsic oscillator) of the segmental oscillator network. If the excitation of both A and B is increased together with a maintained excitability difference, a phase lag between the two oscillators would remain, regardless of the swimcycle duration. Some mechanism of this general type will most likely account for the constant phase lag. There is no evidence to suggest that the oscillator networks along the spinal cord could provide a gradient (Grillner et al., 1986a, 1988a). The inhibitory CC interneurons (Figure 3E)

55

have caudally directed main axons, which form synapses with neurons of at least 20 segments (Buchanan, 1982). In a spinal cord piece of 20 segments, segment 1 will receive inhibition only for the CC neurons of segment 1, whereas segment 20 will receive CC- inhibition from all 20 segments. The degree of CC-inhibition could affect the general excitability of each oscillator and create an excitability gradient along the spinal cord. This would no doubt form a phase lag along the spinal cord, but most likely other features must be added to achieve the invariant 1% phase lag between each segment. An alternative and more plausible possibility may be to assume that oscillators of approximately the same excitability (e.g., resting period length 1.0 s) are connected with weak excitation. If the most rostral oscillator (A) would receive some additional excitation such that it would be active at a period of 0.9 s instead of 1.0 s, then segment B would be entrained and thus also active at a period of 0.95 Hz, but with a phase lag. Each caudal oscillator C, D, E, etc., would be entrained by its rostral neighbour at this higher rate, and between each segment there would be a similar entrainment effect and presumably a similar phase lag (e.g., 1%). The excitatory interneurons excite caudally located neurons over a few segments (Dale, 1986; Buchanan & Grillner, 1987a) and could cause an effect of this type (Matsushima & Grillner, unpublished).

There are as yet simulations based on identified cellular interactions (Grillner et al., 1990), but coupled oscillatory theory has been used to model the phase coupling (Cohen et al., 1982; Ermentrout & Kopell, 1989; Williams et al., 1990). As discussed above, it is likely that GABAergic presynaptic gating mechanisms contribute to the intersegmental coordination (Alford et al., 1990).

4 Equilibrium control during locomotion

The lamprey, as all other vertebrates, maintains its body oriented in relation to gravity and thus swims with its back upwards. The orientation depends on vestibular receptors (Lowenstein et al., 1968) and is abolished if the vestibular apparatus is lesioned (e.g., de Burlet & Versteegh, 1930; Rubinson, 1974). The inputs from the vestibular nerve reach the vestibular nuclei, which project directly to the rostral part of the spinal cord via crossed and uncrossed vestibulospinal projections and indirectly by an activation of reticulospinal neurons (Rovainen, 1979a).

Reticulospinal neurons in different nuclei respond to lateral tilt in a reproducible fashion and to movements in the sagittal plane to a varying degree (nose up-down), as recently shown by Orlosvky et al., (1990). The reticulospinal output signal becomes complex, since reticulospinal neurons have not only vestibular input, but they also forward the locomotor drive signal (see Figure 2) and receive excitatory and inhibitory locomotor-related input from the spinal cord (Kasicki et al., 1988; Dubuc & Grillner, 1989). This latter signal could possibly provide a net gating mechanisms that generate the corrections that maintain body position are, however, as yet unclear. The neuronal machinery in the tectum is probably involved. There are projections from the vestibular nuclei to the tectum (Heier, 1948), in which visual, vestibular, lateral line, and somatosensory inputs are organised in a laminar fashion. The tectum is normally involved in orienting responses.

5 Steering and adaptation to the environment

The purpose of locomotion is normally to move the animal to a different point in space, which obviously requires neuronal machinery for steering and orientation to or from different objects. Consequently, elaborate sensory processing of the surrounding space is needed. Visual receptors are used to analyse the immediate surroundings, and lateral line input is used to sense vibrations in the water and possibly electroreception and chemical gradients (olfaction to find prey or the appropriate river to spawn). To use these types of signals in terms of steering and guidance of movements, the animal must be able to relate the desired direction of movement to the body position. In terms of chemical gradients, the nervous system must integrate over time (is the concentration increasing or decreasing as the movement progresses?).

The laminar organisation of the tectum is probably important in the formation of maps of the surrounding world in lamprey as in other species and to match information from different sensory systems. Orientation movements can be elicited by electrical stimulation of the tectum in the goldfish, and, in the dogfish, clear steering commands to trunk and fins are elicited (Grillner & Wallén, 1984). Preliminary experiments in the lamprey suggest a similar organisation (Ullén, unpublished). The processing mechanisms in the tectum are not yet known. The tecto-reticular projections are probably used for the steering commands, and perhaps the fast Müller cells relay the effects, when rapid control is necessary.

6 Other vertebrate in vitro models

It is pertinent to mention briefly other vertebrate models that have been used. The frog embryo model has been analysed in some detail, and the circuitry underlying the motor pattern consisting of alternating action potentials on the left and right side of the spinal cord is known (Roberts et al., 1986; Roberts & Tunstall, 1990). Great general similarities exist in terms of cellular organisation, transmitters, etc., between lamprey and the frog embryo locomotor systems (Roberts et al., 1986; Sillar & Roberts, 1988).

The in vitro preparations of rat (Kudo & Yamada, 1987; Smith & Feldman, 1987; Cazalet et al., 1990) and chick (O'Donovan, 1986) can generate the motor pattern of locomotion. They should become very useful in the analysis of the basic neuronal mechanisms underlying locomotion in 'higher' vertebrates.

7 Concluding remarks

With regard to the neuronal mechanisms, the transmitter and circuitry used for propulsion in lamprey are largerly known. The backbone of the cellular bases of segmental pattern generation has been unravelled, including both central and sensory mechanisms, as well as the immediate supraspinal mechanisms used in the initiation of locomotion. Simulation has been a necessary aid in this analysis. The intersegmental coordination can be explained in the form of coupled segmental oscillators, but the detailed cellular mechanisms remain to be elucidated. The control systems used for maintaining body orientation, steering and adaptation to the environment during locomotion are now being analysed, but their neuronal bases are not yet understood.

References

Alford, S., Christenson, J., & Grillner, S. (1991). Presynaptic $GABA_A$ and $GABA_B$ receptor-mediated phasic modulation in axons of spinal interneurons. *European Journal of Neuroscience*, **3**, 107-117.

Alford, S., & Grillner, S. (1990). CNQX and DNQX block non-NMDA synaptic transmission but not NMDA evoked fictive locomotion in lamprey spinal cord. *Brain Research*, **506**, 297-302.

Alford, S., & Sigvardt, K. (1989). Excitatory neurotransmission activates voltage-dependent properties in neurons in spinal motor system of lamprey. *Journal of Neurophysiology*, **62**, 334-341.

Alford, S., & Williams, T.L. (1989). Endogenous activation of glycine- and NMDA-receptors in lamprey spinal cord during fictive locomotion. *Journal of Neuroscience*, **9**, 2792-2800.

Andrew, P.C., Pollock, H.G., Elliott, W.M., Youson, J.H., & Plisetskaya, E.M. (1988). Isolation and characterization of a variant somatostatin-14 and two related somatostatins of 34 and 37 residues from lamprey (Petromyzon marinus). *Journal of Biological Chemistry*, **263**, 15809-15814.

Ascher, P., & Nowack, I. (1988). The role of divalent cations in the N-methyl-D-aspartate responses of mouse central neurones in culture. *Journal of Physiology*, **399**, 247-266.

Bainbrigde, R. (1963). Caudal fin and body movements in the propulsion of some fish. *Journal of Experimental Biology*, **40**, 23-56.

Baumgarten, H.G. (1972). Biogenic monoamines in the cyclostome and lower vertebrate brain. *Progress in Histochemistry and Cytochemistry*, **4**, 1-90.

Bongianni, F., Christenson, J., Grillner, S., & Hökfelt, T. (1989). Relation between identified primary sensory neurons and neuropeptide Y- or GABA-immunoreactive neurons and fibers in the spinal cord. *European Journal of Neuroscience*, **73**, 6 (Suppl.)

Brodin, L. (1989). Transmitters, receptors and ionic mechanisms in the control of spinal motor circuits. Physiological and morphological studies of the lamprey CNS with special reference to excitatory amino acid transmission. PhD thesis. Karolinska Inst., Stockholm, Sweden. 169 pp.

Brodin, L., Buchanan, J. T., Hökfelt, T., Grillner, S., Rehfeld, J. F., Frey, P., Verhofstad, A. A., Dockray, & G. J., Walsh, J. H. (1988a). Immunohistochemical studies of cholesystokinin (CCK)-like peptides and their relation to CGRP, 5-HT and bombesin immunoreactivities in the brainstem and spinal cord of lampreys. *Journal of Comparative Neurology*, **271**, 1-18.

Brodin, L., Buchanan, J. T., Hökfelt, T., Grillner, S., & Verhofstad, A. A. (1986). A spinal projection of 5-hydroxytryptamine neurons in the lamprey brainstem. *Neuroscience Letters*, **67**, 53-57.

Brodin, L., Christenson, J., & Grillner, S. (1987). Single sensory neurons activate excitatory amino acid receptors in the lamprey spinal cord. *Neuroscience Letters*, **75**, 75-79.

Brodin, L., Dale, N., Christenson, J., Storm-Mathiesen, J., & Hökfelt, T., Grillner, S. (1990a). Three types of GABA-immunoreactive neurons in the lamprey spinal cord. *Brain Research*, **508**, 172-175.

Brodin, L., & Grillner, S. (1985). The role of putative excitatory amino acid neurotransmitters in the initiation of locomotion in the lamprey spinal cord I. The effects of excitatory amino acid antagonists. *Brain Research*, **360**, 139-148.

Brodin, L., & Grillner, S. (1986). Effects of magnesium on fictive locomotion induced by activation of N-methyl-D-aspartate (NMDA) receptors in the lamprey spinal cord in vitro. *Brain Research*, **380**, 244-252.

Brodin, L., & Grillner, S. (1990). The lamprey CNS - an experimentally amenable model for studies of synaptic interactions and integrative functions. In *In vitro Preparations from Vertebrate*, ed. H. Jansen. New York: Wiley & Sons.

Brodin, L., Grillner, S., Dubuc, R., Ohta, Y., Kasicki, S., & Hökfelt, T. (1988b). Reticulospinal neurons in lamprey: Transmitters, synaptic interaction and their role during locomotion. *Archives Italiennes de Biologie*, **126**, 317-345.

Brodin, L., Grillner, S., & Rovainen, C. M. (1985). N-Methyl-D-aspartate (NMDA), kainate and quisqualate receptors and the generation of fictive locomotion in the lamprey spinal cord. *Brain Research*, **325**, 302-306.

Brodin, L., Rawitch, A., Taylor, T. A., Ohta, Y., Ring, H., Hökfelt, T., Grillner, S., & Terenius, L. (1990b). Multiple forms of pancreatic polypeptide related compounds in the lamprey CNS: partial characterization and immunohistochemical localization in the brainstem and spinal cord. *Journal of Neuroscience*, **9**, 3428-3442.

Buchanan, J. T. (1982). Identification of interneurons with contralateral caudal axons in the lamprey spinal-cord: Synaptic interactions and morphology. *Journal of Neurophysiology*, **47**, 961-975.

Buchanan, J. T., Brodin, L., Dale, N., & Grillner, S. (1987). Reticulospinal neurones activate excitatory amino acid receptors. *Brain Research*, **408**, 321-325.

Buchanan, J. T., & Cohen, A. H. (1982). Activities of identified interneurons, motoneurons and muscle fibers during fictive swimming in the lamprey and effects of reticulospinal and dorsal cell stimulation. *Journal of Neurophysiology*, **42**, 948-960.

Buchanan. J. T., & Grillner, S. (1987a). Newly identified 'glutamate interneurons' and their role in locomotion in the lamprey spinal cord. *Science*, **236**, 312-314.

Buchanan. J. T., & Grillner, S. (1987b). A new class of small inhibitory interneurones in the lamprey spinal cord. *Brain Research*, **438**, 404-407.

Buchanan. J. T., Grillner, S., Cullheim, S., & Risling, M. (1989). Identification of excitatory interneurons contributing to generation of locomotion in lamprey, Structure, pharmacology, and function. *Journal of Neurophysiology*, **62**, 1-11.

Burlet, de, H. M., & Versteegh, C. (1930). Uber bau and funktion des petromyzonlabyrinthes. *Acta Otolaryngologica Suppl.*, **13**, 5-58.

Burrows, M. (1989). Processing of mechanosensory signals in local reflex pathways of the locust. *Journal of Experimental Biology*, **146**, 209-227.

Cattaert, D., El Manira, A., Marchand, A., & Clarac, F. (1990). Central control of the sensory afferent terminals from a leg chordotonal organ in crayfish in vitro preparation. *Neuroscience Letters*, **108**, 81-87.

Cazalet, J. R., Grillner, P., Menard, I., Cremieux, J., & Clarac, F. (1990). Two types of motor rythmes induced by NMDA and amines in an in vitro spinal cord preparation of neonatal rat. *Neuroscience Letters* , **111**, 116-121.

Christenson, J., Alford, S., & Grillner, S. (1989a). Co-localized somatostatin and GABA activate different conductances in intraspinal mechanoreceptor neurons. *31st International Congress of the Physiological Society*, Helsinki, Finland, ed. L. Hirvonen, p. 4424.

Christenson, J., Bohman, A., Lagerbäck, P. Å., & Grillner, S. (1988a). The dorsal cell, one class of primary sensory neurone in the lamprey spinal cord. I. Touch, pressure but no nociception - A physiological study. *Brain Research*, **440**, 1-8.

Christenson, J., Cullheim, S., Grillner, S., & Hökfelt, T. (1990). 5-hydroxytryptamine immunoreactive varicosities in the lamprey spinal cord have no synaptic specialization - an ultrastructural study. *Brain Research*, **512**, 201-209.

Christenson, J., Franck, J., & Grillner, S. (1989b). Increase in endogenous 5-hydroxytryptamine levels modulates the central network underlying locomotion in the lamprey spinal cord. *Neuroscience Letters*, **100**, 188-192.

Christenson, J., & Grillner, S. (1987). Modulation of synaptic transmission in primary sensory neurons in the lamprey spinal cord. *Neuroscience*, **22**, 1022P (Suppl.).

Christenson, J., Lagerbäck, P. Å, & Grillner, S. (1988b). The dorsal cell, one class of primary sensory neurone in the lamprey spinal cord. A light and electron-microscopical study. *Brain Research*, **440**, 9-17.

Cohen, A. H. (1987). The structure and function of the intersegmental coordination system in the lamprey spinal cord. *Journal of Comparative Physiology*, **160A**, 181-193.

Cohen, A. H., & Harris-Warwick, R. M. (1984). Strychnine eliminates alternating motor output during fictive locomotion in the lamprey. *Brain Research*, **293**, 164-167.

Cohen, A. H., Holmes, P. J., & Rand, R. H. (1982). The nature of the coupling between segmental oscillators of the lamprey spinal generator for locomotion: A mathematical model. *Journal of Mathematical Biology*, **13**, 345-369.

Cohen, A. H., Rossignol, S., & Grillner, S., eds (1988). Neural control of rhythmic movements in vertebrates. New York: Wiley. 500 pp.

Cohen, A., & Wallén, P. (1980). The neuronal correlate of locomotion in fish. 'Fictive swimming' induced in an in vitro preparation of the lamprey spinal cord. *Experimental Brain Research*, **41**, 11-18.

Currie, S. N., & Carlsen, R. C. (1987). Modulated vibration-sensitivity of lamprey Mauthner neurons. *Journal of Experimental Biology*, **129**, 41-51.

Dale, N. (1986). Excitatory synaptic drive for swimming mediated by excitatory amino acid receptors in the lamprey. *Journal of Neuroscience*, **6**, 2662-2675.

Dale, N., & Grillner, S. (1986). Dual-component synaptic potentials in the lamprey mediated by excitatory amino acid receptors. *Journal of Neuroscience*, **6**, 2653-2661.

Dongen, van, P. A. M., Grillner, S., & Hökfelt, T. (1986a). 5-Hydroxytryptamine (serotonin) causes a reduction in the afterhyperpolarization following the action potential in lamprey motoneurons and premotor interneurons. *Brain Research*, **366**, 320-325.

Dongen, van, P. A. M., Hökfelt, T., Grillner, S., Verhofstad, A. A. J., & Steinbusch, H. W. M. (1985b). Possible target neurons of 5-hydroxytryptamine fibers in the lamprey spinal cord. *Journal of Comparative Neurology*, **234**, 523-525.

Dongen, van, P. A. M., Hökfelt, T., Grillner, S., Verhofstad, A. A. J., Steinbusch, H. W. M., Cuello, A. C., & Terenius, L. (1985a). Immunohistochemical demonstration of some putative neurotransmitters in the lamprey spinal cord and spinal ganglia. 5-Hydroxytryptamine-, tachykinin-, and neuropeptide Y-immunoreactive neurons and fibers. *Journal of Comparative Neurology*, **234**, 501-522.

Dongen, van, P. A. M., Theodorsson-Norheim, E., Brodin, E., Hökfelt, T., & Grillner, S., et al., (1986b). Immunohistochemical and chromatographical studies of peptides with tachykinin-like immunoreactivity in the central nervous system of the lamprey. *Peptides*, **7**, 297-313.

Dubuc, R., Cableguen, J.-M., & Rossignol, S. (1985). Rhythmic antidromic discharges in cut dorsal roots during locomotion. *Brain Research*, **329**, 375-378.

Dubuc, R., & Grillner, S. (1989). The role of spinal cord inputs in modulating the activity of reticulospinal neurons during fictive locomotion in the lamprey. *Brain Research*, **483**, 196-200.

Dubuc, R., Ohta, Y., & Grillner, S. (1988). *Acta Physiologica Scandinavica*, **134** (suppl. 575), 39.

Duysens, J., & Pearson, K. G. (1980). Inhibition of flexor burst generation by loading ankle extensor muscles in walking cats. *Brain Research*, **187**, 321-332.

Ermentrout, G. B., & Kopell, N. (1989). Mathematical modelling of central pattern generators. In *Cell to Cell Signalling: From Experiments to Theoretical Models*, ed. A. Goldbeter, pp. 89-98. London: Academic. 647 pp.

Garcia-Rill, E., & Skinner, R. D. (1986). The basal ganglia and the mesencephalic locomotor region. See Grillner et al., 1986b, pp. 77-104.

Getting, P. A. (1988). Comparative analysis of invertebrate central pattern generators. See Cohen et al., 1988, pp. 101-127.

Grillner, P., Hill, R., & Grillner, S. (1990). 7-chlorokynurenic acid blocks NMDA receptor induced fictive locomotion in lamprey - Evidence for a physiological role of the glycine site. *Acta Physiologica Scandinavica*, **141**, 131-132.

Grillner, S. (1974). On the generation of locomotion in the spinal dogfish. *Experimental Brain Research*, **20**, 459-470.

Grillner, S. (1981). Control of locomotion in bipeds, tetrapods and fish. In *Handbook of Physiology, Motor Control*, ed. V. Brooks, pp 1179-1236. Bethesda, MD: Am. Physiol. Soc.

Grillner, S. (1985). Neurobiological bases of rhythmic motor acts in vertebrates. *Science*, **228**, 143-149.

Grillner, S., Brodin, L., Sigvardt, K., & Dale, N. (1986a). On the spinal network generating locomotion in lamprey: Transmitters, membrane properties and circuitry. See Grillner et al., 1986, pp. 335-352.

Grillner, S., Buchanan. J. T., & Lansner, A. (1988a). Simulation of the segmental burst generating network for locomotion in lamprey. *Neuroscience Letters*, **89**, 31-35.

Grillner, S., Buchanan. J. T., Wallén P., & Brodin L. (1988b). Neural control of locomotion in lower vertebrates - From behavior to ionic mechanisms. See Cohen et al., 1988, pp 1-40.

Grillner, S., Christenson, J., Brodin L., Wallén P., Hill, R. H., et al., (1989). Locomotor system in lamprey: Neuronal mechanisms controlling spinal rhythm generation. In *Neuronal and Cellular Oscillators*, ed. J. W. Jacklet, pp. 407-434. New York: Dekker.

Grillner, S., & Kashin, S. (1976). On the generation and performance of swimming in fish. In *Neural Control of Locomotion*, ed. R. Herman, S. Grillner, P. Stein, D. Stuart, pp. 181-202. New York: Plenum.

Grillner, S., McClellan, A., & Perret, C. (1981a). Entrainment of the spinal pattern generators for swimming by mechanosensitive elements in the lamprey spinal cord in vitro. *Brain Research*, **217**, 380-386.

Grillner, S., McClellan, A., Sigvardt, K., Wallén, P, & Wilén, M. (1981b). Activation of NMDA receptors elicits 'fictive locomotion' in lamprey spinal cord in vitro. *Acta Physiologica Scandinavica*, **113**, 549-551.

Grillner, S., Perret, C., & Zangger, P. (1976). Central generation of locomotion in the spinal dogfish. *Brain Research*, **109**, 255-269.

Grillner, S., & Rossignol, S. (1978). On the initiation of the swing phase of locomotion in chronic spinal cats. *Brain Research*, **146**, 269-277.

Grillner, S., Stein, P. S. G., Stuart, D., Forssberg, H., & Herman, R., eds. (1986b). *Neurobiology of Vertebrate Locomotion*. London/New York: Macmillan.

Grillner, S., & Wallén, P. (1977). Is there a peripheral control of the central pattern generators for swimming in dogfish? *Brain Research*, **127**, 291-295.

Grillner, S., & Wallén, P. (1980). Does the central pattern generator for locomotion in lamprey depend on glycine inhibition? *Acta Physiologica Scandinavica*, **110**, 103-105.

Grillner, S., & Wallén, P. (1982). On the peripheral control mechanisms acting on the central pattern generators for swimming in the dogfish. *Journal of Experimental Biology*, **98**, 1-22.

Grillner, S., & Wallén, P. (1984). How does the lamprey CNS make the lamprey swim? *Journal of Experimental Biology*, **112**, 337-357.

Grillner, S., & Wallén, P. (1985a). The ionic mechanisms underlying NMDA receptor induced, TTX-resistant membrane potential oscillations in lamprey neurones active during locomotion. *Neuroscience Letters*, **60**, 289-294.

Grillner, S., & Wallén, P. (1985b). Central pattern generators for loomotion, with special reference to vertebrates. *Annual Review of Neuroscience*, **8**, 233-261.

Grillner, S., Wallén, P., Brodin, L., Lansner, A. Ekeberg, Ö., et al. (1990). Neuronal network generating lamprey locomotion - experiments and simulations - supraspinal, intersegmental mechanisms. *Society of Neuroscience Abstracts*, **16**.

Grillner, S., Wallén, P., Dale, N., Brodin, L., Buchanan, J., & Hill, R. (1987). Transmitters, membrane properties and network circuitry in the control of locomotion in lamprey. *Trends in Neurosciences*, **10**, 23-41.

Grillner, S., Williams, T. L., & Lagerbäck, P. Å. (1984). The edge cell, a possible intraspinal mechanoreceptor. *Science*, **223**, 500-503.

Gustafsson, B. (1974). Afterhyperpolarization and the control of repetitive firing in spinal neurones of the cat. *Acta Physiologica Scandinavica*, **416** (Suppl.).

Harris-Warrick, R. M., & Cohen, A. H. (1985). Serotonin modulates the central pattern generator for locomotion in the isolated lamprey spinal cord. *Journal of Experimental Biology*, **116**, 27-46.

60

Harris-Warrick, R. M., McPhee, J. C., & Philler, J. A. (1985). Distribution of serotonergic neurons and processes in the lamprey spinal cord. *Neuroscience*, **14**, 1127-1140.

Heier, P. (1948). Fundamental principles in the structure of the brain. A study of the brain of Petromyzon fluviatiles. *Acta Anatomica*, **8**, 1-213 (Suppl.).

Hill, R. H., Brodin, L., & Grillner, S. (1989). Activation of N-methyl-D-aspartate (NMDA) receptors augments repolarizing responses in lamprey spinal neurons. *Brain Research*, **499**, 388-392.

Homma, S. (1970). Presence of mono-aminergic neurons in the spinal cord and intestine of the lamprey, Lampetra japonica. *Archives of Histology Japonica*, **32**, 383-393.

Homma, S. (1981). Effects of DL-aminoadipate on synaptic transmission in spinal interneurones of the lamprey. *Journal of Comparative Physiology*, **143**, 423-426.

Homma, S., & Rovainen, C. M. (1978). Conductance increases produced by Tγ-aminobutyric acid and glycine in lamprey interneurons. *Journal of Physiology*, **279**, 231-252.

Jordan, L. M. (1986). Initiation of locomotion from the mammalian brainstem. See Grillner et al., 1986b, pp 21-28.

Kahn, J. A. (1982). Patterns of synaptic inhibition in motoneurones and interneurones during fictive swimming in lamprey or revealed by Cl- injections. *Journal of Comparative Physiology*, **147**, 189-194.

Kasicki, S., Grillner, S., Ohta, Y., Dubuc, R., & Brodin, L. (1988). Phasic modulation of reticulospinal neurones during fictive locomotion and other types of motor activity in lamprey. *Brain Research*, **484**, 203-216.

Kudo, N., & Yamada, T. (1987). N-methyl-D-aspartate-induced locomotor activity in a spinal cord - hindlimb muscles preparation of the newborn rat studied in vitro. *Neuroscience Letters*, **75**, 43-48.

Lansner, A., Ekeberg, Ö., Tråvén, h., Brodin, L., Wallén, P., et al.,. (1990). Simulation of the experimentally established segmental supraspinal and sensory circuitry underlying locomotion in lamprey. *European Journal of Neuroscience*, In press.

Lowenstein, O., Osborne, M. P., & Thornhill, R. A. (1968). The anatomy and ultrastructure of the labryrinth of the lamprey (Lampetra fluviatillis L.). *Proceedings of the Royal Society London Series B*, **170**, 113-134.

Mayer, M. L., & Westbrook, G. (1987). The physiology of excitatory amino acids in the vertebrate central nervous system. *Progress in Neurobiology*, **28**, 197-276.

McClellan, A. (1988). Brainstem command systems for locomotion in the lamprey: Localization of descending pathways in the spinal cord. *Brain Research*, **457**, 338-349.

McClellan, A., & Grillner, S. (1983). Initiation and sensory gating of 'fictive swimming' and withdrawal responses in an in vitro preparation of the lamprey spinal cord. *Brain Research*, **269**, 237-250.

McClellan, A., & Grillner, S. (1984). Activation of 'fictive' swimming by electrical microstimulation of brainstem locomotor regions in an in vitro preparation of the lamprey central nervous system. *Brain Research*, **300**, 357-361.

McClellan, A., & Sigvardt, K. (1988.) Features of entrainment of spinal pattern generators for locomotor activity in the lamprey spinal cord. *Journal of Neuroscience*, **8**, 133-145.

Moore, L.E., Hill, R. H., & Grillner, S. (1987). Voltage clamp analysis of lamprey neurons - role of N-methyl-D-aspartate receptors in fictive locomotion. *Brain Research*, **419**, 397-402.

Nieuwenhuys, R. (1977). The lamprey brain in comparative perspective. *Annals of the New York Academy of Sciences*, **299**, 97-145.

Noga, B. R., Kettler, J., & Jordan, L. M. (1988). Locomotion produced in mesencephalic cats by injections of putative transmitter substances and antagonists into the medial reticular formation and the pontomedullary locomotor strip. *Journal of Neuroscience*, **8**, 2074-2086.

Northcutt, R. G. (1984). Evolution of the vertebrate central nervous system: Patterns and processes. *American Zoologist*, **24**, 701-716.

Nowak, L., Bregetovski, P., Ascher, P., Herbet, A., & Prochiantz, A. (1984). Magnesium gates glutamate activated channels in mouse central neurones. *Nature*, **307**, 462-465.

O'Donovan, M. J. (1986). Experimental analysis of motor development in the chick embryo. See Grillner et al., 1986b, pp 415-431.

Ohta, Y., Brodin, L., Grillner, S., Hökfelt, T., & Walsh, J. H. (1988). Possible target neurons of the reticulospinal cholecystokinin (CCK) projections to the lamprey spinal cord: Immunohistochemistry combined with Lucifer yellow staining. *Brain Research*, **445**, 400-403.

Ohta, Y., & Grillner, S. (1989). Monosynaptic excitatory amino acid transmission from the posterior rhombencephalic reticular nucleus to spinal neurones involved in the control of locomotion in lamprey. *Journal of Neurophysiology*, **62**, 1079-1089.

Orlovsky, G. N., Deliagina, T. G., Grillner, S., & Wallén, P. (1990). How does the lamprey maintain an upright body position during swimming? 1. Responses of reticulospinal neurons to natural stimulation of vestibular receptors. *Acta Physiologica Scandinavica*, In press.

Orlovsky, G. N., & Shik, M. L. (1976). Control of locomotion: A neurophysiological analysis of the cat locomotor system. *International Review in Physiology and Neurophysiology*, **10**, 281-317.

Poon, M. (1980). Induction of swimming in lamprey by L-DOPA and amino acids. *Journal of Comparative Physiology*, **136**, 337-344.

Roberts, A., Kahn, J. A., Soffe, S. R., & Clarke, J. D. W. (1981). Neural control of swimming in a vertebrate. *Science* , **213**, 1032-1034.

Roberts, A., Soffe, S. R., & Dale, N. (19860. Spinal interneurons and swimming in frog embryos. See Grillner et al., 1986b, pp. 279-306.

Roberts, A., & Tunstall, M. J. (1990). Mutual re-excitation with post-inhibitory rebound: A simulation study on mechanisms for locomotor rhythm generation in the spinal cord of Xenopus embryos. *European Journal of Neuroscience*, **2**, 11-23.

Ronan, M. (1989). Origins of the descending spinal projections in petromyzontid and myxinoid agnathans. *Journal of Comparative Neurology*, **281**, 54-68.

Rossignol, S., Lund, J. P., & Drew, T. (1988). The role of sensory inputs in regulating patterns of rhythmical movements in higher vertebrates. See Cohen et al., (1988), pp. 201-283.

Rovainen, C. M. (1974a). Synaptic interactions of identified nerve cells in the spinal cord of the sea lamprey. *Journal of Comparative Neurology*, **154**, 189-206.

Rovainen, C. M. (1974b). Synaptic interactions of reticulospinal neurons and nerve cells in the spinal cord of the sea lamprey. *Journal of Comparative Neurology*, **154**, 207-223.

Rovainen, C. M. (1979a). Electrophysiology of vestibulospinal and vestibuloreticulospinal systems in lampreys. *Journal of Neurophysiology*, **42**, 745-766.

Rovainen, C. M. (1979b). Neurobiology of lampreys. *Physiological Reviews*, **59**, 1007-77.

Rubinson, K. (1974). The central distribution of VIIIth nerve afferents in larval Petromyzon marinus. *Brain, Behavior and Evolution*, **10**, 121-129.

Russell, D. F., & Wallén, P. (1983). On the control of myotomal motoneurones during 'fictive swimming' in the lamprey spinal cord in vitro. *Acta Physiologica Scandinavica* , **117**, 161-170.

Selverston, A. I., & Moulins, M. (1985). Oscillatory neural networks. *Annual Reviews of Physiology*, **47**, 29-48.

Shik, M. L., & Orlovsky, G. N. (1976). Neurophysiology of locomotor automatism. *Physiological Reviews*, **56**, 465-501.

Sigvardt, K. A. (1989). Spinal mechanisms in the control of lamprey swimming. *American Zoologist*, **29**, 19-35.

Sigvardt, K. A., & Grillner, S. (1981). Spinal neuronal activity during fictive locomotion in the lamprey. *Society of Neuroscience Abstracts*, **7**,362.

Sigvardt, K. A., Grillner, S., Wallén, P., & Van Dongen, P. A. M. (1985). Activation of NMDA receptors elicits fictive locomotion and bistable membrane properties in the lamprey spinal cord. *Brain Research*, **336**, 390-395.

Sillar, K. T., & Roberts, A. (1988). A neuronal mechanism for sensory gating during locomotion in a vertebrate. *Nature*, **331**, 262-265.

Small, S. A., Kandel, E. R., & Hawkins, R. D. (1989). Activity-dependent enhancement of presynaptic inhibition in Aplysia sensory neurons. *Science*, **243**, 1602-1605.

Smith, J. C., & Feldman, J. L. (1987). In vitro brainstem-spinal cord preparation for study of motor patterns for mammalian respiration and locomotion. *Journal of Neuroscience Methods*, **21**, 321-333.

Stein, P. S. G. (1978). Motor systems, with special reference to the control of locomotion. *Annual Review of Neuroscience*, **1**, 61-81.

Viana di Prisco, G., Wallén P., & Grillner, S. (1989). Spinal target neurons and synaptic effects of sensory neurons responsible for entrainment of central locomotory pattern generators in lamprey. *31st International Congress of the Physiological Society* , Helsinki, Finland, p. 4459.

Viana di Prisco, G., Wallén P., & Grillner, S. (1990). Synaptic effects of intraspinal stretch receptor neurons mediating movement-related feedback during locomotion. *Brain Research*, **530**, 161-166.

Wallén P., Buchanan. J. T., Grillner, S., Christenson, J., & Hökfelt, T. (1989). The effects of 5-hydroxytryptamine on the afterhyperpolarization, spike frequency regulation and oscillatory membrane properties in lamprey spinal cord neurons. *Journal of Neurophysiology*, **61**, 759-768.

Wallén P., & Grillner, S., (1987). N-methyl-D-aspartate receptor induced, inherent oscillatory activity in neurons active during ficitive locomotion in the lamprey. *Journal of Neuroscience*, **7**, 2745-2755.

Wallén P., Grillner, S., Feldman, J., & Bergelt, S. (1985). Dorsal and ventral myotome motoneurons and their input during fictive locomotion in lamprey. *Journal of Neuroscience*, **5**, 654-651.

Wallén P., & Lansner, A. (1984). Do the motoneurones constitute a part of the spinal network generating the swimming rhythm in the lamprey. *Journal of Experimental Biology*, **113**, 493-497.

Wallén P., & Williams, T. L. (1984). Fictive locomotion in the lamprey spinal cord in vitro compared with swimming in the intact and spinal animal. *Journal of Physiology*, **347**, 225-239.

Watkins, J. C., & Evans, A. H. (1981). Excitatory amino acids transmitters. *Annual Review of Pharmacology and Toxicology*, **21**, 165-204.

Webb, P. W., Kostecki, P. T., & Stevens, E. D. (1984). The effect of size and swimming speed on locomotor kinematics of rainbow trout. *Journal of Experimental Biology*, **109**, 77-95.

Wickelgren, W. O. (1977). Physiological and anatomical characteristics of reticulospinal neurones in lamprey. *Journal of Physiology*, **270**, 89-114.

Williams, T. L., Grillner, S., Smoljaninov, V. V., Wallén P., Kashin, S., & Rossignol, S. (1989). Locomotion in lamprey and trout: The relative timing of activation and movement. *Journal of Experimental Biology*, **143**, 559-566.

Williams, T. L., Sigvardt, K. A., Kopell, N., Ermentrout, G. B., & Remler, M. P. (1990). Forcing of coupled nonlinear oscillators: Studies of intersequential coordination in the lamprey locomotor central pattern generator. *Journal of Neurophysiology*, **64**, 862-871.

Chapter 6: The brain in motor action: New approaches to probe its functioning

F. Lopes da Silva

The current knowledge on how coordinated motor control programs are functionally organised in the human brain is being investigated using different, and complementary approaches. Different disciplines (neurology, neurophysiology, neuroanatomy, computer sciences) and methodologies have given a significant contribution to obtain an insight into this question.
In this chapter, first, a few elementary concepts from the theory of neural networks and of control systems are discussed in relation to the organisation of motor functions.
Second, recent data on the neurophysiology of the cortical neuronal populations responsible for high level motor programs are illustrated.
Third, the new methodologies that are now being used to obtain information from the living human brain regarding the functional organisation of motor programs are presented, with the emphasis on the following: event-related cross correlation, movement-related magnetic fields, position emission tomography (PET) and related techniques.

1 Introduction

In the last decade, our knowledge about the physiology of the processes responsible for motor control has advanced considerably in several respects. First, an insight into these processes, at the *theoretical* level, has been gained through the development of computational models. Second, a finer analysis of the *anatomical* circuits involved in motor control has revealed how different brain areas are concatenated in order for complex movements to occur. Third, knowledge of the *electrophysiology of cortical neuronal populations* engaged in the preparation and execution of movements has led to a better understanding of how motor programs are represented in the neuronal networks of the cortex. Fourth, quantitative *brain imaging techniques*, namely using electroencephalographic and magnetoencephalographic recordings associated to other brain imaging methods, made possible the analysis in real time of the functional organisation of complex movements in the awake human brain.

Here a brief overview of the processes responsible for the brain control of coordinated motor activity in man is presented. As theme for this overview, the visually guided reaching and prehension movement of the hand was chosen, due to the fact that there is a wealth of experimental results and theoretical models available regarding this type of behaviour.

2 Theoretical aspects

A number of *theoretical aspects*, derived from the theory of control systems, are relevant for the analysis of motor behaviour. In general, we may consider that the brain systems involved in this and other types of behaviour, can be viewed as a set of interrelated *functional units*. Such an unit is called a *motor schema* by Arbib (1989). In this overview, we will rely on the terminology developed by Arbib (1989) which offers a good basis for linking theoretical concepts with anatomo-functional notions. A motor act, for instance reaching and grasping an object, results from the coordinate activation of a number of motor functional units or schemas. In the terminology of computer science, this activation can be represented by the execution of a *program* 'advance the hand' that must stop as the object to be grasped is reached. Therefore, the execution of the program must continue under the control of a single test: 'is the object reached?'. This control implies that *feedback* is necessary. In this way, the subject compares the intended result with the actual situation, and computes a compensatory signal as necessary. An analysis of the behaviour considered here, i.e. reaching and grasping

Author's address: Vakgroep Experimentele Dierkunde, Kruislaan 320, 1098 SM Amsterdam, The Netherlands.

movement of the hand, reveals that there are two motor functional units at work, one that controls the arm movement and an other that controls the preshaping of the hand, i.e., the adjustment of the fingers and their correct orientation. In addition, there are perception functional units which task is to gather information from the environment through vision and touch. The perceptual schemas assemble information about (i) the location of the object to be seized (target), (ii) its size, shape and (iii) its orientation.

An analysis of this behaviour leads to the conclusion that the movement is under the control of systems with different dynamics: an initial *fast stage* that controls the ballistic movement of the arm (feedforward), and a final *slow stage* that uses feedback for continual sensing of the deviation between the real hand position and that of the target. The movement of the fingers can be considered as being accomplished in the form of the movements of three sets of fingers (or virtual fingers = VF): one is the thumb (VF1), another is constituted by two, or three, fingers to grasp the object (VF2) and the other by the remaining fingers that help to support the operation (VF3). Accordingly, we assume that there are different brain programs to control the movement of the three sets VF1,2,3 (Arbib, 1989).

Our interest is to establish how a given behaviour is realised in terms of neuronal activity. In this respect, an important question is to relate the behavioural operations to the activity of the responsible brain areas. In general, we may consider this problem in terms of two parallel lines: the structural and the functional line. Functionally, we may consider that a given *behaviour* is realised due to the coherent activity of *functional units* (or *schemas* in the sense of Arbib's) that consist of related groups of neurons forming *distributed arrays of neuronal networks* subserving the same function. In terms of structure, these distributed networks occupy specific *brain regions,* that need not be contiguous but that are interconnected by long- and short-range connections, and within which *layers, columns* and/or *modules* formed by concatenated groups of *neurons,* can be distinguished.

3 Anatomical aspects

The main brain regions, which contain those functional units of neuronal networks that are involved in the realisation of the different motor programs, indicated above, have been identified by different experimental methods. In first instance, through the analysis of the results of lesions and also through anatomical tracer, studies aimed at identifying the circuitry of the different areas. We may point out that lesions can cause distinct effects depending on the brain area affected. For instance, lesions of the parietal cortex (parietal area 5 or PA5) will result in spatiomotor discoordination and misreaching, of the pre-frontal and pre-motor cortices (dorsal and postarcuate premotor area, PMd and PMa, and PM, see also below) in deficits in learning of motor strategies and of the motor cortex (MI) in paralysis. This indicates that different motor areas are involved in distinct levels of motor programming. Indeed, we have to realise that there are several *motor systems.*

Of these many systems we will consider here only those that form the *functional units* responsible for the execution of the motor programs described above. In this respect, an important set of systems are formed by the *arrays of neuronal networks* of the *somato-motor cortex.* This nomenclature emphasises the fact that the motor and somatosensory cortex are strongly interconnected. It is a common characteristic of these areas that they are strongly interconnected and that several re-entrant circuits between these areas exist. In fact, the idea that the planning and control of movements would occur as the result of the simple serial, or hierarchical, activation of distinct brain areas can not be sustained on the light of modern neurobiological knowledge.

Anatomically, the brain region of central interest for the topic of this review can be subdivided into the primary somatosensory cortex (S1) located in the anterior parietal lobe and the motor cortical area 4 lying more rostrally. A general scheme of the localisation of these and other different cortical areas is shown in Figure 1. Within the former region, we can distinguish four cytoarchitectonic areas: area 3a, 3b, 1 and 2. *Area 3a* processes proprioceptive information from the joints and muscle spindles, and provides feedback signals for the precise positioning of the joints. *Area 3b* processes cutaneous information from slowly and rapidly adapting receptors with small receptive fields. *Area 1* combines inputs from area 3b, that supply information from cutaneous receptors, including skin vibration, with those from area 3a that inform about the speed of movement, in order to perform texture analysis. *Area 2* receives inputs from the other areas mentioned above so that the corresponding neurons have large receptive fields and respond to complex features of stimuli such as the direction of a stimulus during exploratory touch. Areas 2, 1 and 3a have re-entrant connections with the *motor cortical area (4 or MI)*. MI is the primary cortical area that is topographically organised in relation to the body musculature. Strick and Preston (1982 a,b) have identified in this motor cortical area separated *kinaesthetic and tactile maps*, the latter lying more caudally than the former.

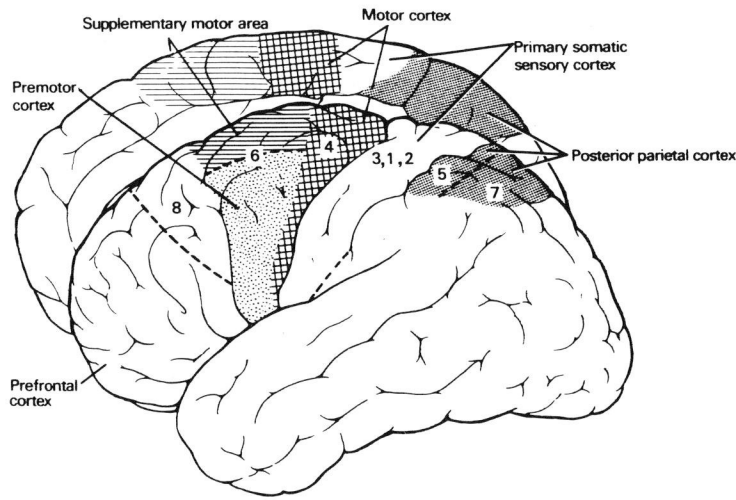

Figure 1. Main cortical areas of the human brain responsible for motor control. The numbers representing the cytoarchitectonic fields according to the classic nomenclature of Brodmann are indicated (Adapted from Kandel & Schwartz, 1985, p. 495).

We can now relate the analysis of the movement presented theoretically above with the functional organisation of the somato-motor cortex.

Firstly, we consider the preshaping of the hand and the formation of the virtual fingers. The former takes place mainly under proprioceptive control where the *functional units* of area 3a play a major role. The control signal that gives information about enclosing the object with the hand, when contact is established, is most likely processed in areas 1 and 2. The finer movements of prehension and adjustment are guided by functional units of the tactile map of MI. The constitution of the virtual fingers (VF) is the task of the kinaesthetic motor map of MI or Area 4. How the command signals that initiate the *preshaping* and the movement of *enclosing* of the object by the hand are generated is discussed below.

Secondly, we consider the component of the behaviour that we may describe in terms of the *command to 'advance the hand and grasp the object'*. This command is most likely generated outside the motor cortex. Indeed, other brain regions (including the basal ganglia and the cerebellum, which will be left out of this discussion) participate, in addition to the somatomotor cortex, in the realisation of this behaviour. According to the hypothesis of Iberall and Arbib (1989), the motivational input that leads to the execution of the program 'advance the hand and grasp the object' will activate functional units of area 7. The interplay of area 7 with Areas 8 and 9 (frontal eye fields) results in the activation of the perceptual *functional units* responsible for finding the location of the object and the desired orientation, *i.e., the approach vector selector*. In parallel herewith, the premotor area (PM) that cues for the *schemas* "orient trunk and move arm" would also be activated.

Thirdly, we consider how the commands that activate the *preshape schema* and the *enclose schema* are generated according to the theory put forward by Iberall and Arbib (1989), which is schematically illustrated in Figure 2. Both *schemas* are mediated by distributed arrays of neuronal networks, in which we should include the supplementary motor area (SMA), the dorsal and postarcuate premotor areas (PMd and PMa, collectively denoted as arcuate premotor area or APA), PA5 (parietal cortex area 5), and all the somato-motor cortex areas described above (MI or area 4, areas 2, 1, 3a and 3b). The arcuate premotor cortical areas compute the desired preshape state of the hand, based on signals about the wrist position. SMA controls the preshaping by gating inputs until the wrist is correctly positioned and then releases inhibition to enable the closing of the fingers using information about the virtual fingers from areas 5 and the kinaesthetic map of area 4.

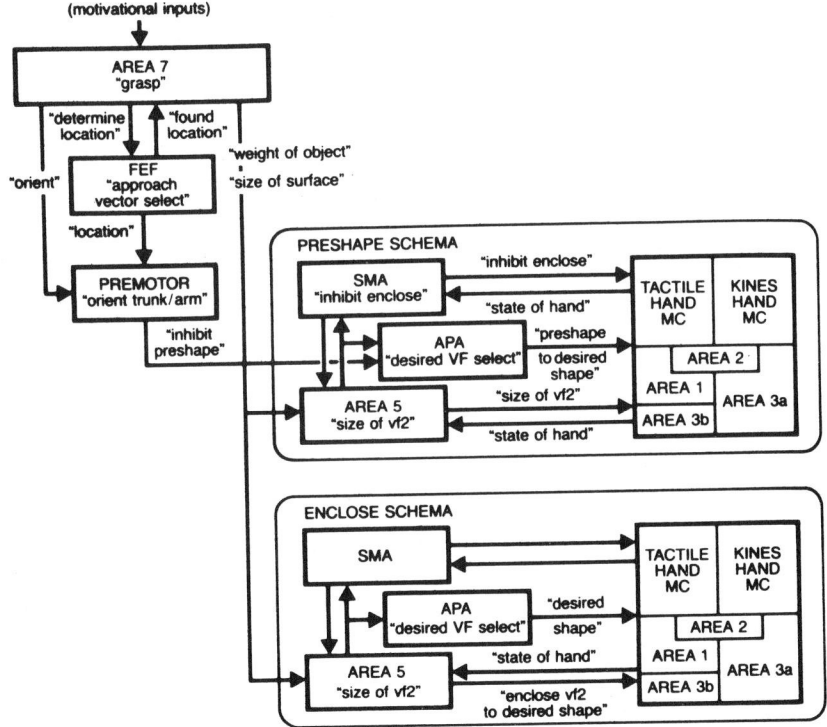

Figure 2. Scheme representing the theory of Iberall and Arbib for the coordinated motor program of the reaching and hand grasping behaviour. (Adapted from Arbib, 1989).

In this way, the stage is set for the coordinated motor actions that are essential for programming the desired movement of the hand. The control of this movement is realised by way of signals that inform the cortex about the current spatial state of the hand in terms of virtual fingers. These signals are provided by areas 1 and 2. An essential aspect of this *control system* is the comparison between the desired and the actual states of the hand. This comparison may be achieved by an operation of subtraction such as realised in control systems. In this way, a *command directional signal* would be created as the output of the motor cortex that enables the desired movement.

4 Electrophysiology of cortical neuronal populations

The neural activity of three cortical areas MI, PA5 and PMd has been the most studied, in order to find neurophysiological signals that encode for the different stages of motor behaviour such as described here. The most used experimental animal has been the monkey. Georgopoulos (1990; Georgopoulos et al., 1983, 1986) recorded the activity of neurons in the motor cortex, namely MI area, of rhesus monkeys on the contralateral side to the arm that had to move to a visually guided target (Figure 3). They found that a large proportion of neurons discharged according to the angle by which the movement of the arm departed from the preferred direction. In addition, the vector sum of the preferred directions of each neuron weighted by the corresponding firing rates yields a *population vector*, that is a good predictor of the intentional direction of movement. The length of the population vector is a good predictor of velocity. Therefore, this neuronal population exhibits *directional tuning*.

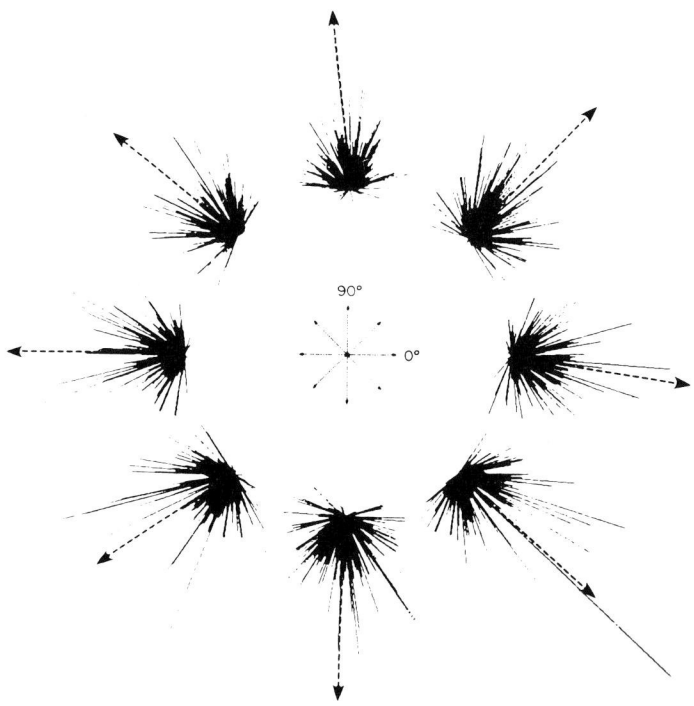

Figure 3. Illustration of the activity of a neuronal population of the motor cortex of monkey, consisting of 241 neurons, coding the direction of reaching movements. Eight movement directions are indicated in the central figure. For each cell, the preferred direction is indicated by a solid vector. The resulting vector for the whole population is shown as an interrupted line. Note the close agreement between the direction of the movement and that of the population vector (Adapted from Georgopoulos et al., 1983).

One question of general interest in the frame of the present discussion, is whether the neurophysiological signals recorded in these and similar experiments, reflect the *kinematics* of the movement, i.e., the trajectories and the actual physical characteristics of the movement itself (direction, force, velocity, acceleration) or its dynamics, i.e., the time evolution of the changes due to different forces. This was tested by letting the animal make a movement with the same trajectory but under different loads, i.e., under conditions of variable dynamics. In this respect, the behaviour of the different cortical areas is not uniform. Studies of Kalaska and collaborators (Kalaska, 1991; Kalaska et al., 1990) have shown that the neurons most sensitive to load were found in area MI, which has an important projection to the motoneurons of the spinal cord by way of the corticospinal pathway. Therefore, it appears that the neuronal populations of MI are important in coding the dynamic characteristics of the movement. At the other extreme, the neurons of PA5 were little sensitive to loads but strongly correlated with arm trajectory, i.e., with the representation of the directional vector as such. Those of PMd lay in-between those of the two other areas. Also regional differences between different cortical areas were put in evidence by comparing the neuronal activity during the time of movement preparation with that occurring during execution. Experimentally, this was studied by introducing a time delay between the time at which the command signal was given and the initiation of the overt movement (Crammond & Kalaska, 1989; Kalaska & Crammond, 1992). During the delay period, many neurons displaying directional tuning sensitivity were found; most of them in PMd, less in PA5 and much less in MI. Many neurons were active both during the preparatory and during the execution phases. However, the majority of MI neurons were active only during movement execution.

An interesting feature of the behaviour of these neuronal populations is that their discharge patterns also depend on the context in which the movement takes place. Different patterns of firing of cortical neurons, particularly those outside area MI, may be found depending on whether the movement was executed to reach food or to remove an aversive stimulus (Kalaska & Crammond, 1992). This reinforces the idea that the organisation of the neuronal activity responsible for motor control does not obey the principle of a fixed chain of commands but shows flexibility and can be adjusted to changing environmental situations.

5 Neuronal network models

The insight acquired with the electrophysiological experiments described above has led to recent attempts to simulate the behaviour of the cortical neuronal populations by network models such as that of Jeannerod (1991) where a network of elementary neurons interconnected in a multilayer structure is simulated. In this and other similar models (Kalaska & Crammond, 1992), the neuronal elements have activity tuned to a particular direction. An input signal that gives the indication of the direction of the movement to be executed is projected onto the multilayer network. Additional signals represent information about appropriate reference frames. The resultant signal from the whole network is computed according to the principles of network self-organisation depending on weighted connections between the neuronal elements and on learning through several trials: 'learning by doing' (Loeb, 1983).

6 Brain imaging techniques

The possibility of gathering physiological information about the neuronal processes involved in coordinated movements *non-invasively in human* was appreciably enhanced in recent years by new technical developments. *Electrophysiological recordings* have led to the identification of a number of movement-related potentials that can be recorded from the intact scalp, namely: the *readiness or Bereitschaftspotential* (BP), the *motor potential* (MP), the *premotion positivity* (PMP) and *the reafferent potential* (RAP) (Deecke & Kornhuber, 1987). Several studies on

69

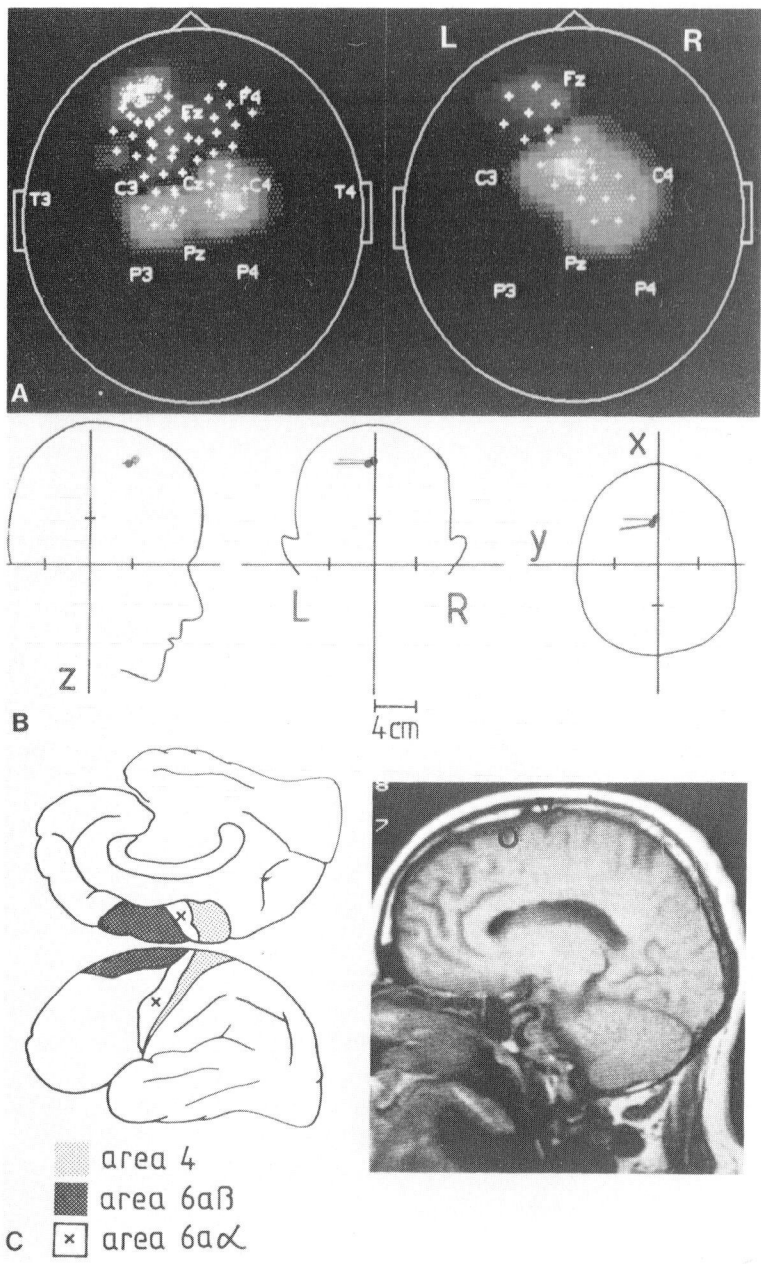

Figure 4. A. - Brain maps of magnetic flux densities taken from the interval 1.2 and 1.0 s prior to movement onset; data from two sessions. The crosses indicate the sites at the scalp where the magnetic field was recorded. B - Functional localisation of equivalent current dipole sources corresponding to the maps shown in A. C - Location of SMA in the superior frontal gyrus corresponding to the cytoarchitectonic areas 6ab and 6aa on the lateral and medial surface of the hemisphere; cortical motor area 4 (MI) is also indicated. D - Projection of the equivalent current dipole on a sagittal section through the left mesial cortex (circle). (Adapted from Lang et al., 1991).

the topography of these different movement-related potentials have been carried out with the aim of relating these electrophysiological phenomena with the activity of different brain areas. For instance, it was noted that the BP can be sub-divided into two components: the first one starts about 1.2 s prior to movement onset and is bilaterally and symmetrically distributed with a maximum recorded above the SMA; the second component starts later, about 0.5 s prior to movement onset and is asymmetrically distributed with a maximum above the contralateral MI area in finger movements. Using *magnetic field recordings* in a patient with an asymmetrical lesion of the region around the SMA, it was possible (Lang et al., 1991) to consistently localise a current dipole source in the intact SMA starting about 1.2 s prior to the initiation of a voluntary movement of the thumb. In addition, it was also demonstrated that a current dipole source in the intact MI area was able to account for the second component of BP at 0.6 s prior to movement onset (Figure 4). Furthermore, other non-invasive methods have contributed to our knowledge of the topographic distribution of different neuronal populations involved in movement control, namely measurements of blood flow and metabolic parameters using isotopes combined with computerised tomography such as in positron emission topography (PET) and related methods.

7 Conclusions

In this overview, I have chosen a particular type of complex movement, the reaching and prehension movement of arm and hand, in order to illustrate how this type of movement control can be analysed at different levels. The ultimate aim of such an analysis is, of course, to integrate the anatomy, the physiology and the behaviour into a comprehensive theoretical model that accounts for the essential properties of the movement. This example demonstrates the necessity of following parallel research lines in order to reach this ultimate goal. In this respect, it is interesting to underline the importance of the new data obtained using electrophysiological recordings of populations of cortical neurons which have led to a better insight into how the representation of movement programs takes place in the cortex. Furthermore, the advances in brain imaging techniques begin to offer new possibilities of realising in real time an analysis of the functional organisation of the human brain in relation to the control of movement.

Acknowledgements. I wish to thank Trúc Ngô-Hà and Cristine Knaap-Cabi for their secretarial assistance.

References

Arbib, M.A. (1989). *The Metaphorical Brain 2*, (pp. 458).New York: Wiley.

Crammond, D.J., & Kalaska, J.F. (1989). Neuronal activity in primate parietal cortex area 5 varies with intended movement direction during an instructed delay period. *Experimental Brain Research*, **76**, 458-462.

Deecke, L., Groezinger, B., & Kornhuber, H.H. (1987). Voluntary finger movement in man: cerebral potentials and theory. *Biological Cybernetics*, **23**, 99-119.

Georgopoulos, A.P., Caminiti, R., Kalaska, J.F., & Massey, J.T. (1983). Spatial coding of movement. A hypothesis concerning the coding of movement direction by motor cortical populations. In J. Massion, J. Paillard, W. Schulz and M. Wiesendanger (Eds.), *Neural coding of motor performance* (pp 327-336). Springer-Verlag.

Georgopoulos, A.P., Schwartz, A.B., & Kettner, R.E. (1986). Neuronal population coding of movement direction. *Science*, **233**, 1416-1419.

Georgopoulos, A.P. (1990). Neural coding of the direction of reaching and a comparison with saccadic eye movements. In: Cold Spring Harbor Symposia on Quantitative Biology, vol. LV: The Brain, (pp. 849-860).

Iberall, T., & Arbib, M.A. (1989). Schemas for the control of hand movements. An essay on cortical localization. Quoted in Arbib (1989), "The metaphorical Brain 2".(p. 458). New York: Wiley

Jeannerod, M. (1991). In J. Paillard (Ed.), *Brain and Space* (pp. 49-69). Oxford: Oxford University Press.

Kalaska, J.F., Cohen, D.A.D., Prud'homme, M., & Hyde, M.L. (1990). Parietal area 5 neuronal activity encodes movement kinematics, not movement dynamics. *Experimental Brain Research,* **80**, 351-364.

Kalaska, J.F. (1991).What parameters of reaching are encoded by discharges of cortical cells. In D.R. Humphrey and H.-J. Freund (Eds.), *Motor control: concepts and issues* (pp. 307-330). Chichester, U.K: Wiley.

Kalaska, J.F., & Crammond, D.J. (1992). Cerebral cortical mechanisms of reaching movements. *Science,* **255**, 1517-1523.

Kandel, E.R., & Schwartz, J.H. (Eds.) (1985). *Principles of neural science* , 2nd edition. Amsterdam: Elsevier.

Lang, W., Cheyne, D., Kristeva, R., Beisteiner, R., Lindinger, G., & Deecke L. (1991). Three-dimensional localization of SMA activity preceding voluntary movement. *Experimental Brain Research,* **87**, 688-695.

Loeb, G.E. (1983). Finding common ground between robotics and physiology. *Trends In Neurosciences,* **6**, 203.

Strick, P.L., & Preston, J.B. (1982a). Two representations of the hand in area 4 of a primate. 1. Motor output organization. *Journal of Neurophysiology*, **48**, 139-149.

Strick, P.L., & Preston, J.B. (1982b). Two representations of the hand in area 4 of a primate. 2. Somatosensory input organization. *Journal of Neurophysiology*, **48**, 150-159.

Chapter 7: Some reflections upon the work presented on the control of locomotion in a primitive vertebrate

E. Otten

Grillner's main motive to try and understand the control system of locomotion in a primitive vertebrate is that neural nets in higher vertebrates are too complex and inaccessible. This is indeed the case: There is quite a gap between simulation of neural networks and the functional analyses of them in terms of anatomical and physiological measurements. However, by selecting a particular function, in this case locomotion, of a primitive vertebrate, problems that may have been solved in later stages of evolution are not studied. For instance the stunning observations in sensory-motor integration in human subjects have as little to do with locomotion of a lamprey as a learning robot playing the piano has to do with a punched paper cylinder driven pianola.

On the other hand, it is very useful to simulate an existing neural network together with its neurochemistry. Since neural networks by virtue of their combinatorial power can reproduce almost any pattern, their simulation on a computer has very little explanatory power. By providing the boundary conditions given by the observations of a network in vivo, such simulations may have much more in common with the actual process going on.

If one has any experience in running simulations of complex processes, it is clear that fine tuning of a model is an art, that can result in realistic looking output. The realism of the output is not a proof of the completeness or correct setting of parameters of the model. Fine tuning can cover up incompleteness. Therefore such a model, if fine tuning is used, forms a very coherent, reproducible, and transferable theory, but is not more than a theory. This is an essence of numerical simulation in biological and medical sciences.

It should be pointed out that there is nothing truly innovative in the research presented by Grillner. However, as is the case with biological systems themselves, the power of the research presented is the combination and interaction between the parts. Doing histochemistry of neurons is nothing new, but using the data in computer simulation and from there go back to the living preparation is something that is not done very often and in that sense this kind of research can provide us with the contours of a general theory of motor control.

Author's address: Department of Neurobiology and Oral Physiology, Bloemsingel 10, 9712 KZ Groningen, The Netherlands.

Chapter 8: A word about the work presented on methods to probe brain function

E. Otten

The roots of any theory about the world and the beings that inhabit it lie in observations. Observations however are limited by the bare sense organs with which the researcher is equipped. Most modern publications on motor control show results of observations done with equipment using all sorts of transducers. The development of electronics enables the researcher to record from intact human beings and display the results in a form that is accessible to the unaided sense organs. Although these representations often have a beauty of themselves (this is very likely not a coincidence), one should not forget that they are not more than a set of observations. These observations do not contain any structure of a theory. As was indicated by a number of philosophers of science in the fifties and sixties, what we see depends largely on what we know. Images do not speak for themselves, but resonate with our knowledge. One can temporarily be blinded by the colourful display given by Lopes da Silva on modern techniques of positron (not position!) emission tomography, recordings of movement related magnetic fields etc., but this state should rapidly be transformed to one of speculation on the process and origin of these images.

Neurophysiologists find the work of their heroes mainly in the American and British scientific publications. If one visits the labs where these heroes work, one finds a striking contrast: TheAmericanlabs are crammed with the latest computers and recording equipment and most of the time people discuss ways of getting the best image of some process, while on the British Islands experimental conditions are made possible by the presence of rubber bands and adhesive tape. This contrast has a lot to do with the difference in financial conditions in which research has to be performed, but it is certainly also a reflection of a difference in scientific paradigm. It is a British tradition to spend a large part of scientific effort in constructing a coherent theory, while the American tradition is in getting results across (think of the unsurpassed American museums). In summarising the contribution of Lopes da Silva, it becomes clear that he has a strong affiliation to the latter. His compelling presence in the Dutch scientific community has certainly been fruitful in this way. Working in the Netherlands may also implicate that the British heritage of formulating theories invites him to start integrating these beautiful images into an explicit theory of motor control.

Author's address: Department of Neurobiology and Oral Physiology, Bloemsingel 10, 9712 KZ Groningen, The Netherlands.

Chapter 9: What's new in motor control?

C.C.A.M. Gielen

One of the achievements in motor control research in the past few years is that gradually a synthesis has become available of various aspects of the motor control system in man and animal. In the early years spinalised and anaesthetised animals have been used as a model to study some aspects of motor control. Although this type of research was necessary to explore the main characteristics and the enormous complexity of the motor system, it has led to some artefacts. For example the reflex studies on isolated soleus muscle preparations were quite successful in revealing major properties of the various reflex components. However, they also have contributed to the notion that reflex actions are restricted to a single muscle that is stretched and they have completely ignored that reflexes are highly coordinated responses of the motor system to a wide variety of perturbations (see e.g., Matthews, 1990).

At the other hand, most studies on human motor control have considered the motor system as a black box, the properties of which have been studied with various behavioural and system analysis techniques. These studies provide a lot of information on the functional properties of the human motor system. However, it is hard and frequently impossible to establish a relation between functional properties and neuronal structures.

One of the major achievements made in the last few years is that new techniques have provided the tools to bridge the gap between functional properties and neurophysiological data. With new experimental techniques such as PET-scan, MRI, magnetic cortical stimulation, etc., it has become possible to do experiments in behaving animals and in man as to the neuronal structures that are involved in various motor tasks, during a wide variety of motor activity.

In addition to the experimental techniques new theoretical techniques have been developed which provide the tools to model the motor system. These techniques frequently use *a system theoretical approach*, which describes the hierarchy of various functional elements in the motor system, combined with methods such as *neural networks*, which provide the tools to describe the neuronal information processing by neurons within a functional element. The latter technique gives an opportunity to 'open the black box'. The major contributions of the theoretical approaches are that 1) they provide a synthesis between various functional elements in the motor system into an integrated scheme including various types of interactions and with feedback/feedforward loops, and 2) the new models not only describe the available data into a coherent model, but also deliver specific hypotheses that can be tested experimentally. The latter is certainly necessary if one tries to *understand* the motor system, rather than building an encyclopaedia with facts.

The presentations by dr. Lopes da Silva en dr. Grillner have illustrated the usefulness of a combined approach with traditional and new experimental and theoretical techniques. As a result, the interdisciplinary collaboration will certainly benefit from these developments since the new techniques allow a synthesis of approaches at various conceptual levels, thereby providing a bridge between disciplines such as psychology, neurophysiology, biology, and physics.

References

Matthews, P.B.C. (1990). The knee jerk: still an enigma? *Canadian Journal of Physiology and Pharmacology*, **68**, 347-354.

Author's addres: Department of Medical Physics and Biophysics, University of Nijmegen, Geert Grooteplein Noord 21, 6525 EZ Nijmegen, The Netherlands.

Chapter 10: CPG or CIA?

Onno G. Meijer[1], Auke A. Post[1] & Rob Bongaardt[1,2]

1 Flexibility in headless frogs

During the 1848 revolution, Eduard Friedrich Wilhelm Pflüger (1829-1920) was still in his adolescence. At the time, his great dream was one united, democratic Germany, and in 1849 he was taken prisoner, threatened with execution. He escaped and found refuge in the physiology of the spinal frog. In 1853 his book on *The Functions of the Spinal Cord* appeared.

> When a pair [of copulating frogs] is captured, and one only grips the male, the female will be pulled out of the water as well because the male strongly holds [her] enclosed … . When the spinal cord of the male is cut between the atlas and the second vertebra … he does not let go … . If … she attempts to withdraw, he just holds her stronger … . If one now drops some acetic acid on one of his arms, he lets go with that arm, whereas the other one holds the female, and he rubs off the biting substance with the ipsilateral leg. Then, however, he again holds her enclosed with both arms … . *(Pflüger, 1853, pp. 17-18, our translation.)*

In Pflüger's experiments, it was clear that the neurologically (not: literally) headless frog is able to remove acid from its body with one of its legs. Pflüger then cut off that very leg. Upon renewed irritation with acid, the stump moved aimlessly for a while, and then another leg was used to remove the acid (cf. Figure 1). What better evidence is there for the intelligence of the headless frog? Pflüger concluded that there must be a spinal soul, thereby suggesting that the source of intelligence is divisible, much to the dismay of his contemporaries (e.g., Lotze, 1853).

Twentieth-century movement sciences have abolished the soul altogether. So, Pflüger's phraseology has been discarded, but still, his experiments continue to amaze the scientific community (Meijer et al., 1988). To date, it is known that the frog-with-a-head will jump in the air and shudder upon being irritated, and that the headless frog will restrict its actions to the affected side (Fukson et al., 1980). The fact remains, however, that the headless frog behaves with more functional flexibility then one would have expected. Evidently, something is wrong with our expectations.

The present paper has not been written to explain Pflüger's experiments. His work is used as the first stepping stone for an argument that shows, or at least aims to show, what the authors regard as a deep, persistent problem in modelling motor coordination and control. The functional nature of motor coordination and control invites the researcher to choose between different types of model. It will be argued that choosing one type of model will highlight one important aspect of coordination and control, but obfuscate the other.

In the opinion of the present authors, Pflüger's frogs exemplify one important aspect of coordination and control, i.e., flexibility. Functional *flexibility* implies being able to achieve a particular goal ('functionality'), no matter—within limits—what happens on the way. Other labels could be used, but the phenomenon is important.

Authors' address: 1) Departement for the Theory and History of Human Movement Sciences, Faculty of Human Movement Sciences, Vrije Universiteit, Van der Boechorststraat 9, 1081 BT, Amsterdam, The Netherlands, and 2) Unit for Theoretical Psychology, Faculty of Psychology and Pedagogics, Vrije Universiteit, De Boelelaan 1111, 1081 HV, Amsterdam, The Netherlands.

We summarise with stepping stone 1:

> *Flexibly functional movement behaviour is not dependent, or at least not entirely dependent, upon the brain.*

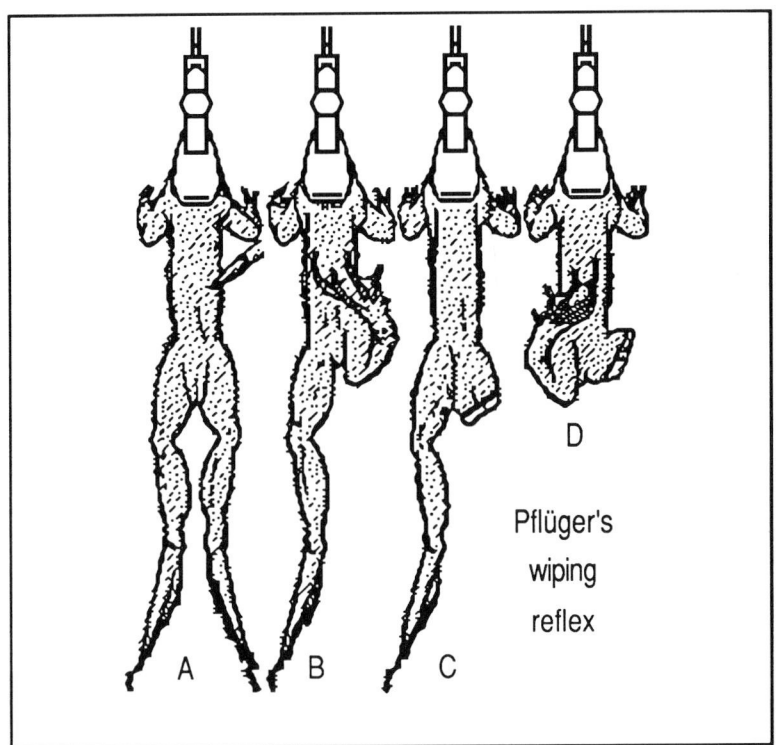

Figure 1. Pflüger's wiping reflex. A. The headless frog is irritated with acid on the right side of its back. B. The irritant is removed with the ipsilateral hind-paw. C. That paw is then cut off. D. After renewed irritation, the stump moves aimlessly for some time, and then the acid is removed with the other hind-paw. (Digitised image adapted from Verworn, 1907, p. 198.)

2 Anticipation in spinal cats

Pflüger thought in terms of a *spinal* soul, an idea which was a democratically inspired rejection of the conviction that all intelligence has to come from the highest level.

Changing the emphasis, one may also observe that Pflüger thought in terms of a spinal *soul*, thereby implicitly accepting the idea that for any form of intelligent behaviour a specific *agency* must be responsible[1]. From such a viewpoint, functional movement behaviour depends on a central intelligence agency, if not in the head, then in the spinal cord. Accordingly, the scientific community started to focus on the spinal cord per se, rather than the locomotor apparatus or cooperation between the environment, the nervous system, and the locomotor apparatus.

---

[1] One is reminded of Dennett's (1979) 'loan of intelligence', i.e., 'explaining' intelligence by invoking an intelligent agency. Eventually, such a 'loan' must be repaid, that is to say, explained as a property of the underlying organisation of the system.

In 1894, the physiologist Friedländer cut an earthworm in two, sewed the parts together and saw the reunited segments move in a coordinated fashion (cf. Von Holst, 1937). Friedländer concluded that anatomical continuity of the cord was not essential for movement. Apparently, he argued, movement coordination depends on *reflex chains*, one segment affecting the other, just as in toppling dominoes. To Friedländer, coordination was not a purely spinal phenomenon: In the perspective of the present paper, his thinking was in terms of cooperation.

As far as we know, it was McDougall (1903) who first claimed that 'automatic' activities *within the central nervous system* are (co-)responsible for rhythmic behaviour. Also Brown (1914) explicitly attacked the reflex theory. He anaesthetised cats and observed that complex rhythmic movements occur at a depth of narcosis where reflexes are abolished. Brown suggested that rhythmic movements are "conditioned by the balanced activation of antagonistic and linked efferent cycles (or half centres)" (1914, p. 36), exerting mutual inhibition. In other words, Brown envisaged movement to depend on central agencies which function more or less automatically: They intermittently activate and silence each other.

In 1931, Adrian and Buytendijk recorded oscillatory brainstem activity in goldfish. This oscillatory activity was shown to correlate with respiration. In 1937, Von Holst performed experiments on deafferented and isolated fish spinal cords, and concluded that these were able to generate rhythmic activity[2].

In 1961, Wilson published a paper on the flight of the locust. He had worked with deafferented preparations, or with preparations that consisted of only the head, the part of the cord containing the somata for the thoracic nerve, and a ventral cuticular strip. In these preparations, patterned activity could still be seen, 'similar' to the pattern in normal flight albeit with a lower frequency. In order for the pattern to emerge, Wilson argued, sensory information is not needed: Locust flight is essentially under central control. He was careful enough, however, to add that sensory information came in to modulate the pattern. But it would act on top of what was already available in the central nervous system.

In 1969, Engberg and Lundberg reported their experiments on what they described as 'cooperative' cats. Engberg and Lundberg performed EMG analyses of the pattern of muscle contraction in the hindlimb during 'unrestrained locomotion'. They observed that the main activity of the extensor muscles starts shortly before the foot touches the ground. Physiologically, this contraction cannot be timed by foot pressure reflexes or changes in joint position due to the gravitational torque after touching the ground, since the contraction precedes it. Functionally, then, the contraction of the extensors is of an anticipatory nature.

There appears to be nothing magical in the 'anticipatory' contraction of extensor muscles in the locomoting spinal cat. It may, for instance, be triggered by receptors in the connective tissues of the leg, or the timing may be dependent on the organisation of the spinal cord.

The present authors, however, would like to suggest that a problem is hidden behind this apparent simplicity, i.e., how it is that biological systems are able to 'anticipate'. Recently, Robert Rosen (1985) developed a formal theory of 'anticipatory systems' as *systems that use predictive models to control their present behaviour*. We regard Rosen's initiative as a very important one. According to Rosen, any theory of life ought to imply a theory of anticipation (cf. Rosen, 1991). There exist, for instance, biochemical cascades where chemical concentrations early in the cascade (co-)determine the concentrations of enzymes that are used later in the cascade. Nervous systems, to take another example, allow for the embodiment of predictive models. Much of social behaviour, Rosen argues, can only

---

[2] This should not be taken to suggest that Adrian and Buytendijk, or Von Holst, had a centralist conception of motor control. Both the Cambridge zoology group and Von Holst played an important role in recognising the importance of sensory input.

be understood if one takes anticipation into account.

There are certainly disadvantages to invoking 'predictive models'. They are reminiscent of 'intelligent agencies' which so often just beg the question. On the other hand, including 'predictive models' in the analysis of living systems, allows for linking final and causal styles of reasoning, both so typical of biological science. Anticipation also allows for errors to appear (such as when something goes wrong halfway a chemical cascade), and thus for evolution and learning.

We propose to regard extensor timing in Engberg and Lundberg's locomoting spinal cats as an instance of 'anticipation' in Rosen's sense, and thus, of the use of a predictive model.

We summarise with stepping stone 2:

Anticipatory control, i.e., being co-controlled by a predictive model, has been observed in spinal animals.

3 Central pattern generators

From the end of the sixties onwards, vertebrate researchers took the spinal cord to exert anticipatory control[3] over the locomotor apparatus. This development was stimulated by a line of research focusing on the spinal 'programming' of rhythmic movements.

In a short but epoch making paper, Grillner and Zangger (1975) wondered "How detailed is the central pattern generation for locomotion?" The authors had studied hindlimb electromyograms of deafferented acute mesencephalic cats. They first remark that the EMG-pattern and kinematic parameters of the acute mesencephalic cat, able to walk on a treadmill, "appear identical to that of the walking intact cat" (p. 367). It is then stated that upon deafferentation "the pattern appears unchanged" (p. 369), which leads to the following conclusion:

> the central program (unknown neuronal design) does not simply generate an alternate activation of flexors and extensors but a more delicate pattern that will sequentially start and terminate the activity in the appropriate muscles at the correct instance. The role of afferents in this context is not primarily to control the timing of the muscle activity in the individual limb but may rather be to interact when external perturbations occur. (*p. 370.*)

Mainly through the work of the Grillner group and of Selverston and his co-workers (Grillner & Zangger, 1979; Selverston, 1980), it became popular to write about *central pattern generators* (CPGs), almost as if these were anatomically recognisable devices, ideally surrounded by a membrane, and functioning as music boxes: Switch them on and the kinetic melodies will be played. It was as if a scientific revolution had occurred and a certain pride to partake in such a revolution can be seen in the literature from that period. Stated Calabrese:

> In 1879 T.H. Huxley made an analogy between the rhythmic movements of crayfish in response to sensory stimulation and the melody which issues from a music box when the stop is pressed. He concluded, "It is in the ganglia that we must look for the analogue of the musical box. A single impulse conveyed by a sensory nerve to a ganglion, may give rise to a single muscular contraction, but more commonly it originates a series of such, combined to a definite end." This statement, more than any other epitomizes to me the sense of wonder which has driven research on invertebrate CPGs over the past twenty years. (*1980, p. 542.*)

---

[3] The idea of 'anticipatory control' had come from the Engberg and Lundberg (1969) study. Rosen's (1985) formal theory of 'anticipation' was published much later.

As pertains to the average 'scientific revolution' (Beek & Meijer, 1988), however, no definitive research program has emerged yet, and the leading concepts of CPG research have remained extremely vague. 'Central' appears to suggest 'non-sensory', but time and time again authors emphasised the importance of sensory information (cf. Grillner, 1985). 'Pattern' seems to point to something invariant, but variance continued to be faithfully reported (cf. Grillner & Zangger, 1975) and, thus, it never became clear what exactly 'the' pattern was. The notion of 'generator' implies a 'program' (cf. the above Grillner & Zangger quote), but at the same time concepts were used from stability theory, e.g., 'perturbations', notwithstanding the fact that stability theories have been claimed to be incompatible with program theories (Pattee, 1977).

Nevertheless, it was clear that the isolated spinal cord produces 'something' so similar to what one would expect normal activity in the intact animal to be, that it had to be more than coincidental. In an elegant study, Cohen and Wallén (1980), for instance, were able to show that, under the appropriate conditions, the isolated lamprey spinal cord produces patterned activity remarkably similar to what one would expect such activity to be in the intact swimming lamprey (Cohen & Wallén, 1980; Figure 2). Cohen and Wallén coined this activity 'fictive swimming', and regarded it as a neural *correlate* to actual behaviour. The research community at large, however, appeared to be somewhat less cautious, and took this finding as (another) piece of evidence that suggested the spinal cord to be *responsible* for the production of patterned movement.

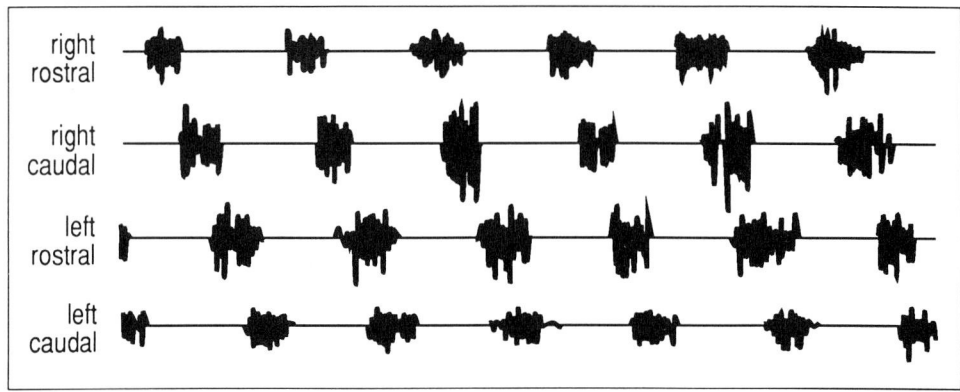

Figure 2. Patterned activity in the isolated lamprey spinal cord. Recordings are given from the right and left ventral roots of the 7th ('rostral') and the 19th ('caudal') segments of an isolated chunk, consisting of 27 segments. (Adapted from Cohen & Wallén, 1980, p. 13.)

The idea of a central pattern generator fits well with the notion of 'anticipation', but when taken to the extreme of a music box, it becomes patently absurd. Apart, maybe, from the above overzealous Calabrese quote, modern researchers have generally refrained from taking the music box metaphor too seriously.

Already in the thirties, Bernstein (1935) had shown that no movement is ever repeated in exactly the same way (Figure 3). The number of music boxes one would require for all possible movements would be more than astronomical. On the other hand, Bernstein also argued that at least something has to remain the same if one is to move functionally at all, and it is very plausible that 'something' has a correlate in the organisation of the nervous system.

Closer inspection of Bernstein's cinematographic analysis of hammering, leads to the conclusion that nothing *metric* remains the same. There is, however, a clear *functional* invariance, i.e., the nail is hit[4]. The question now is: What aspect of the nervous system correlates with that functional invariance? Just for the sake of argument, let us assume there to be a CPG for hammering. Then, there is a macroscopic[5] 1:1 relation between whether or not the organism is hammering and whether or not that CPG is active. At a microscopic level, however, relationships are generally of a non-linear nature (Bernstein, 1935), and no 1:1 relation is to be expected between detailed changes in CPG activity and detailed changes in muscle fibre activity or the metrics of the movement[6].

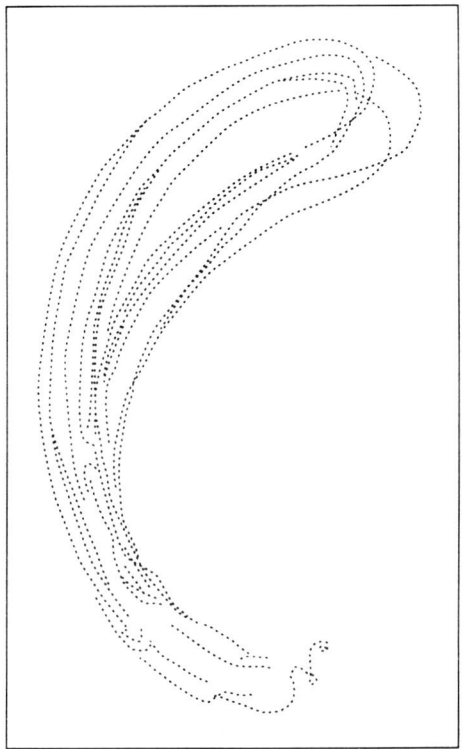

Figure 3. No movement is ever repeated in exactly the same way. Bernstein's cinematographic analysis of a hammering movement in an industrial setting. (Bernstein, 1935, p. 90.)

---

[4] One may note that this functional invariance necessarily coincides with metric variance: Continuing to hit the nail on the head at least implies that the hammer moves deeper with every repetition.

[5] The notion 'macroscopic', here, is used in the sense of general systems theory, i.e., referring to a high level of analysis. Whether or not one actually can refrain from using a microscope is, thus, entirely irrelevant.

[6] In the lamprey, Ayers et al. (1983) compared micro-electrode recordings of *in vitro* patterns of the spinal cord with filmed actual swimming. The isolated cord activity was found to be no closer to swimming than to any other front-to-rear lateral undulation (like burrowing or crawling). Wallén and Williams (1984), on the other hand, compared micro-electrode measurements *in vitro* and *in vivo*. Their conclusion was that "the patterns of activity and silence in the ventral roots of the in vitro preparation are indistinguishable from such patterns in the body musculature of the symmetrically swimming animal" (p. 236). Clearly, the Ayers et al. study is in agreement with what we state, whereas the Wallén and Williams study is not. We blame the lamprey. That is to say, animals which usually behave in a homogeneous, unchanging surround, may learn to cope with the non-linearities just as the guitar player learns to cope with the unpredictability of the strings, i.e., by tuning. In such cases, non-linearities are still hidden in the microscopy, and we would, therefore, regard the resulting invariances as being mesoscopic rather than microscopic.

Maybe CPGs find themselves at some mesoscopic scale. Maybe, then, 1:1 relationships can be found between the global topology of the movement (or the global activity of whole muscles or muscle groups), and the global topology of CPG activity. This is certainly what CPG research appears to suggest (cf. Figure 2). To date, however, even careful analysis of the actual topology has continued to confront us with variance (e.g., Van Soest, 1992). It took Bernstein a lifetime to pinpoint mesoscopic invariances. Apparently, CPG researchers wrestle with the same problem. It is our hunch that the situation is not much better for the moving organisms themselves, when learning to coordinate and control in an always-changing surround.

The above analysis can be taken to illustrate that the starting point of one's research constrains the set of possible outcomes.

In CPG research, muscle fibres are usually looked at in a top-down direction: They are there to be controlled. From the viewpoint of the individual muscle fibre, the anticipatory nature of a particular contraction cannot derive from the muscle fibre itself since, by definition, no single system is able to outrun itself (Rosen, 1974). The anticipatory signal, then, must come from outside the muscle fibre, the CPG being the most attractive candidate to date. In that viewpoint, *CPGs embody the (anticipatory) programs which ensure correct timing*.

There is, however, another way to look. There appear to exist 1:1 relationships between functional states of the animal (e.g., swimming, walking) and the activation state of particular spinal networks (CPGs). In that viewpoint, *CPGs are functional states of parts of the nervous system*. Then, in order to ensure functional stability under changing conditions, flexibility is required. But now we are looking in a bottom-up direction: Working our way from the variance of CPG activity up to the invariance of the overall functional state.

It appears to be difficult, if not impossible, to look in two directions at the same time. Nevertheless, we argue, this is exactly what CPGs appear to require, 'go-betweens' as they are between the overall functional state of the animal, and the specific timing of its muscle fibres. Accordingly, leading authors in the field, such as Grillner, have, in our opinion, continued to present bistable positions: downplaying as well as emphasising the role of sensory information; looking for invariance, looking for variance; thinking in terms of 'programs' and in terms of 'stability'.

We don't blame the research community for having developed what we regard as oscillating between two positions, although one could have been much more explicit about it. Oscillating between positions is the hallmark of paradoxes (Varela, 1979). Accordingly, we want to argue that the problem resides in the paradoxical nature of biological function, given the constraints of scientific modelling. The more one focuses on anticipation, the more problematic it becomes to understand functional flexibility.

We summarise with stepping stone 3:
Understanding central pattern generators as embodiments of predictive models is certainly attractive, but may lead to losing functional flexibility out of sight.

4 Dynamical systems

There is a remarkable stability in biological movement, both in its macroscopic functionality (being able to reach your goal 'no matter' what happens on the way) and in the underlying organisation (nervous rhythms going on 'no matter' how they are disturbed).

If a system is, within limits, able to deal with external perturbations to then return to its original state, that system is, within these limits, *stable*. Modern stability theory derives from the mathematical theory of dynamical systems (cf. Arnol'd, 1983) that occupies itself with iterated mapping of a space upon itself. If the mapping of state x of a dynamical system is a function of that same state, the system in question is said to be *autonomous*. Is one to model autonomous changes over time of a real system, the differential equation will be of the form:

$$\dot{x} = f(x).$$

External influences to the system have to be modelled as functions of time t— time being the only variable the system 'shares' with the outer world. In classical music box models, changes of the locomotor apparatus (or of movements) were regarded as a function of t only: The to-be-controlled was understood as being completely passive in its relationship with the controller. Bernstein's research has shown such modelling to be incorrect. On the other hand, purely autonomous systems don't undergo any external influences at all, and also that situation, we argue, would be highly implausible for any biological system. Hence, we generally have to do with 'mixed systems' (Meijer & Bongaardt, 1992), i.e., systems which change as a function of their own state as well as time:

$$\dot{x} = g(x,t).$$

At the end of the sixties, several elaborated, stochastic forms of the theory of dynamical systems were shown to have rather general validity for the natural sciences, i.e., in understanding and analysing the nature of *macroscopic* processes (Haken, 1983; cf. Prigogine & Stengers, 1984). Examples were presented of macroscopic stability, which could be modelled as (dynamical) autonomy: Given a range of external conditions, the systems in question could be shown to organise their own stable macroscopy. Strictly speaking, these systems are not 'autonomous', since different external conditions will stochastically drive them through a transition into a new form of behaviour. Neither, however, are they completely passive in their relationship with the external conditions. Systems which change autonomously within specific ranges of external conditions, have been called *self-organising systems* (Haken & Wunderlin, 1990)[7].

In self-organising systems, a *high-dimensional* microscopy (such as air molecules in a tornado) 'self-simplifies'—to use Pattee's 1973 term—under the appropriate energy conditions into a *low-dimensional* macroscopy (such as the overall topology of the tornado). The relationship between microscopy and macroscopy is of a dual (circular) nature (Haken & Wunderlin, 1990): Whereas the macroscopic order *emerges* out of the microscopy, at the same time it *forces* the microscopy to behave in concert.

It is of relevance to note that this is essentially how Bernstein envisaged the organisation of motor actions. Bernstein (1957) coined the term 'coordination-control', which can be understood as 'coordination and control'. In terms of the present argument, Bernstein's 'coordination' is the transformation of high into low dimensionality, whereas 'control' is the transformation of low into high dimensionality. There is a certain circularity in their relationship. Starting with a large number of microscopic events at simple threshold devices (e.g., sensors), 'coordination' will lead to a high-level low-dimensional simplification of those events, whereas 'control' starts with such a high-level low-dimensional description which then flows out into the low-level high dimensionality of the locomotor apparatus (e.g., motor units).

Given that no two movements are ever executed in exactly the same way, that there is a fantastic number of degrees of freedom to be controlled, and that control relationships in the living are of a non-linear nature (Bernstein, 1940), the theory of dynamical systems has

---

[7] The term 'self-organisation' has been widely misused (e.g., Jantsch, 1985). However, the above definition is, in our opinion, sufficiently clear and precise. Those who want to avoid the confusion in the general literature, may consider 'partial autonomy'.

remarkable advantages in understanding the organisation of motor actions. Since stability implies flexibility, the theory of dynamical systems can serve to describe and help us to understand functional flexibility.

On the other hand, self-organising systems (and autonomous systems) cannot anticipate, and in order to incorporate anticipation into dynamical models of coordination and control, one will need to add an external timer (a function of time) to the otherwise autonomous differential equation. And thereby, maybe, lose one's original intention: There is now an anticipatory program to co-control a flexible self-organising system. Mathematically, this is not impossible[8], but, again, we are confronted with a situation where one has to look into two directions at the same time.

In a 1980 debate in neurophysiology[9], Selverston wondered "Are central pattern generators understandable?" and expressed himself to be pessimistic: it would be impossible to pinpoint all the relevant anatomical *detail*. Others were more optimistic and emphasised the *emergent properties* of network-organisations (e.g., Davis, 1980). Explicitly, the Selverston (1980) discussion was about searching for more detail versus modelling emergent properties.

By 1980, 'command neurons' were deemed responsible for the switching on and off of neural 'music boxes' (cf. Kupferman & Weiss, 1978). Accordingly, Selverston appears to have reasoned that all that had to be done was to analyse the detailed anatomy in order to know how the melody is played: Selverston's reasoning is closely analogous to program theories. 'Emergent properties', on the other hand, are a hallmark of applying the theory of dynamical systems. Hence, a divide emerged. As far as we know, Avis Cohen (1980) was the first in trying to bridge this divide:

> we must stop thinking in terms of dichotomies between patterns generated as a result of the "emergent properties" of networks and those generated as a result of the "endogenous properties" of cells. (*p. 543*.)

A line of research resulted in which, at least in principle, the theory of dynamical systems was linked to relevant details of the microscopy[10]. After analysing the dynamics of oscillator *coupling*, (e.g., Rand et al., 1988), the nature of anatomical connections between cells could be studied (Figure 4). In a 1990 paper, Ermentrout and Kopell showed that strong *all-to-all coupling* may be pathological (as, for instance, in epileptic seizures; cf. Babloyantz & Destexhe, 1986). The mathematical predictions of *nearest-neighbour coupling* (Williams et al., 1990) turned out to be more consonant with real network behaviour, but nearest-neighbour coupling is not the only connectivity pattern found in microscopic anatomy, and lacks the necessary stability (Kiemel, 1990). Kiemel (1990) showed short-distance coupling from rostral to caudal, in combination with long-distance coupling from caudal to rostral—*Kiemel's coupling*— to be a realistic possibility, that is: in terms of actual network behaviour,

---

[8] Nor is it technically impossible to include non-linear procedures and functions into motor programs (Hinton, 1984).

[9] Over the last decade, the theory of dynamical systems has had a major impact on the *psychology* of motor control (e.g., Haken et al., 1985; Kugler & Turvey, 1987; Beek, 1989). It has been claimed that the notion of 'self-organisation' should be the starting point for any science of biological movement: Dynamical stability reveals, at least, *what* the animal does (Meijer et al., 1988). On the other hand, psychologists like to explain how we are able to anticipate. So, a divide has emerged (Meijer & Roth, 1988) between motor psychologists who (continue to) think in terms of 'programs' (anticipation), and those who think in terms of 'self-organizing systems' (functional flexibility).

[10] The notion of 'self-organisation' has kept a remarkably low profile amongst neuroscientists. Nevertheless, the theory of dynamical systems has gained considerable importance. This was visible from the 1980 Selverston debate onwards.

anatomical data, and stability.

The present authors want to argue that there is, indeed, some closing of the gap in the above understanding of coupling schemes. But the researchers involved would be the first to emphasise that this is still not enough.

In Bernsteinian terms, neural *coordination* has been modelled far better than before. But the problem of *control* remains. In terms of the present argument, the more one focuses on functional flexibility, the more problematic it becomes to understand anticipation.

Stepping stone 4, therefore, mirrors stepping stone 3:
Understanding the functional flexibility of CPGs in terms of the theory of dynamical systems is certainly essential, but may lead to losing anticipation out of sight.

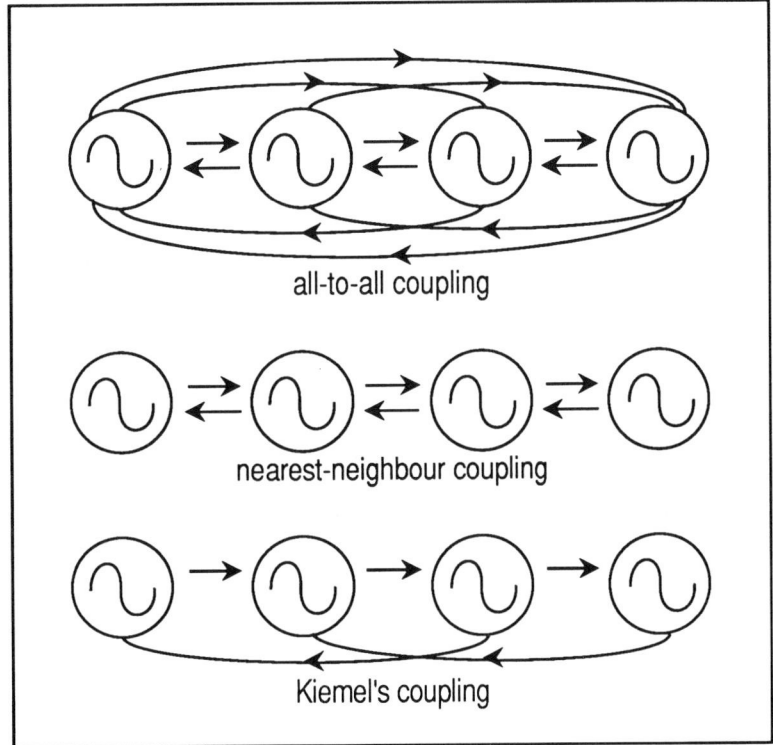

Figure 4. Coupling schemes (left is rostral, right is caudal).

5 The paradox

We have shown functional flexibility (as in Pflüger's frogs) and anticipation (as in Engberg & Lundberg's cats) to be properties of brainless animals. It is clear that spinal organisation may offer important insights into the nature of these two aspects of biological intelligence. We have presented CPGs as important candidates for both. Functional flexibility may be modelled by using the theory of dynamical systems, anticipatory control coming in as an extra function (of time). Anticipation can be modelled by using 'programs', flexibility coming in by invoking IF ... THEN loops with flexible procedures and functions. Where, then, is our deep problem?

85

In so far as a system behaves as programmed, it presents itself in a *symbolic* mode. (Say "Walk!" and it walks.) In so far as a system behaves as autonomous, it presents itself in a *dynamic* mode. ($\dot{x} = f(x)$.) Whenever a system does both, we would like our model to incorporate both, for instance:

$$\dot{x} - f(x) = walk,$$

but mathematically, such a combination cannot make sense since symbols and dynamics are *incompatible* (Pattee, 1977). (The above formalism would read: the changes over time of x, minus a function of X, equals the product of w, a, l, and k—there being no way to have the formalism understand the meaning of 'walk'.) In other words, as soon as the very nature of observables becomes different in principle (such as: kinetic properties of atoms and temperature), it becomes impossible to meaningfully incorporate them into one single model.

The present authors propose that the two ways of looking at CPGs (the programmers of muscle fibres, and the functional states that correspond to the functional behaviours of animals), are instantiations of two styles of reasoning, i.e., symbolic and dynamic, which ultimately rely on observables that are of a different nature in principle. To the best of our knowledge, then, the two ways of looking at CPGs are incompatible. Since it is also clear that we need both ways of looking, we have arrived at a paradox.

Symbols and dynamics are not only incompatible, they are also *equivalent*—just as the two arrows connecting microscopy and macroscopy in the theory of dynamical systems, or Bernstein's notions of 'coordination' and 'control'.

If we accept a system to consist of a set of elements together with their relationships, then any system can be represented both as a *tree* (symbols) and as a *network* (dynamics). Intuitively, this theorem can be understood as follows (Varela, 1979): Whichever relationship one starts with in a network, is equivalent to the stem of a tree; The set of relationships of the next elements are then the first branches, etc. (Figure 5): The network is equivalent to as many infinite trees as there are nodes in the network.

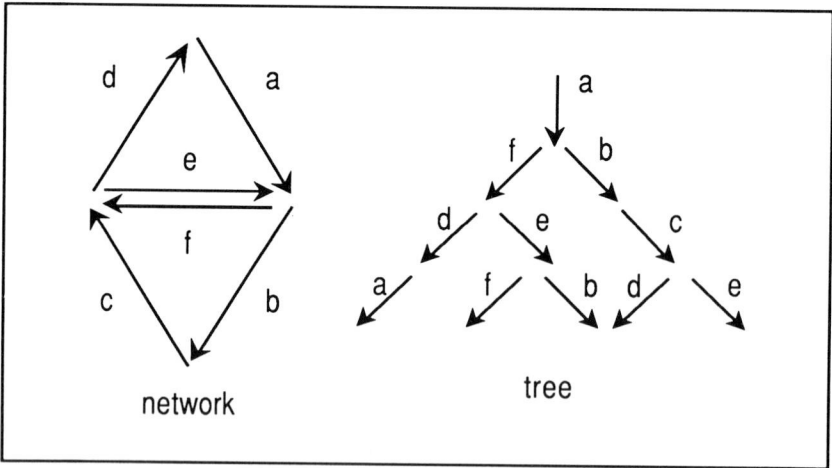

Figure 5. A network is equivalent to a set of infinite trees. (Adapted from Varela, 1979, p. 92.)

This theorem mirrors a self-evident truth in movement science (Meijer et al., 1988): logically, any observed sequence of movement trajectories can be produced by a computer program, and physically, any observed sequence of movement trajectories must be in accordance with the laws of dynamics. The paradox, then, does not arise out of the object of our science, or the nature of ordinary language (in which, at least, we can express there to be a paradox). The paradox has to do with the nature of mathematical modelling. It has been known for a long time that any paradox-free mathematics does not allow for mixing levels. Russell's paradox summarises the argument: The set of sets that don't contain themselves, would contain itself if, and only if it were not to contain itself (cf. Russell, 1979)[11]. Nervous systems, however, appear to be less restrained, and to happily mix levels all the time.

Pattee (1977) has shown that paradox-free mathematical models cannot capture the symbolic and the dynamic mode at the same time. Ordinary language, however, is able to do so without any problem. Apparently, nervous systems are in this respect more akin to ordinary language than to paradox-free mathematics. Several authors have tried to develop a mathematics of paradoxes to provide for that fact (cf. Varela, 1979). Their attempts, however, have not been without problems. At present, the choice is ours: patching up mathematics or investigating networks.

We summarise with stepping stone 5:

No paradox-free mathematics can meaningfully capture functional flexibility and anticipation in the same formalism.

6 State of the art

We know, without the slightest doubt, that there are chunks of nervous tissue that, in isolation under the appropriate conditions, will produce patterned activity so similar to what one would expect the normal rhythm to be, that it is tantalising to conclude that these patterns play an essential role in normal motor action. Anatomically, the detailed structure of these chunks has proven to be rather elusive.

Invertebrates have shown themselves to be attractive research objects: Their nervous systems contain relatively few neurons, many of which are large and individually recognisable. Even in invertebrates, however, the task is far from trivial. It took about ten years to chart the relevant network for *Tritonia* swimming (Getting, 1988). To date, we have reached only a simplified scheme of the swimming CPG in *Tritonia* (Figure 6).

The swimming CPG in *Tritonia* does not have one specific pattern only. It consists of 12 neurons, clustered in three groups. Dorsal and ventral swim interneurons (DSI and VSI) exert reciprocal inhibition. Brain cell C2, excited by DSI, gives delayed excitation to VSI, by which it is inhibited. A tonic drive is needed to sustain oscillatory activity. The six DSIs are mutually connected by excitatory and inhibitory connections, the VSIs by excitatory connections. In rest, the inhibitory DSI connections are dominant; During escape swimming, the excitatory connections take over (Getting & Dekin, 1985).

The general impression one gains, is that of *distributed* coordination and control: There is no single cell or cell cluster on which all details depend.

---

[11] One may want to argue that Russell's paradox has been formulated in set theory and is, therefore, not necessarily valid for the theory of dynamical systems. The theory of dynamical systems, we argue to the contrary, is not only about dynamics, it is also about systems. And systems are sets of elements together with their relationships. Since systems are sets, if set theory has a paradox, systems theory has to follow suit. The relationships between the elements find themselves at a higher 'level', i.e., the level of the system as a whole. Although we have to acknowledge that the notion of 'level' is mathematically vague, we take the paradox to imply that one cannot develop meaningful models encompassing both the microscopy and the macroscopy of the same system.

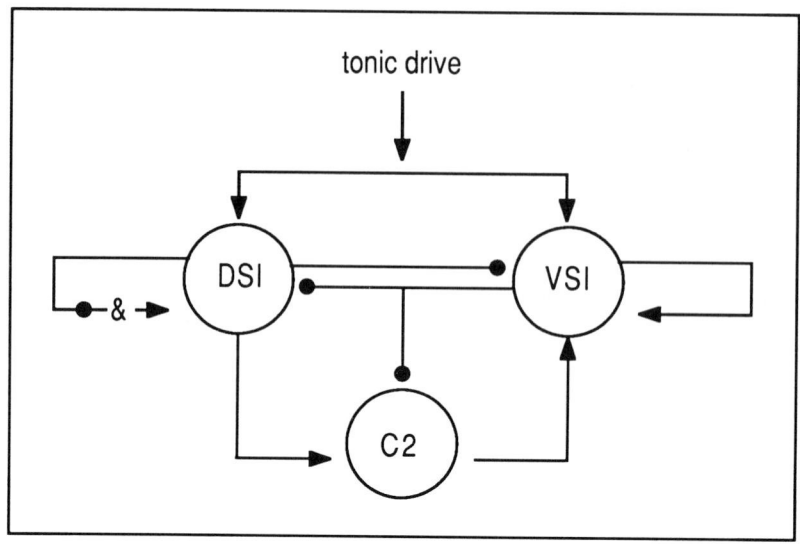

Figure 6. Simplified scheme of the Tritonia swimming CPG. DSI/VSI: dorsal/ventral swim interneurons; C2: cerebral cell 2; arrow: excitation; stick: inhibition. (Adapted from: Getting, 1988, p. 114.)

Similar findings have come from the work on cockroach flight by the Camhi group (Libersat & Camhi, 1988; Libersat et al., 1989). Whenever the sensors 'warn' the cockroach that it is time to fly, the dorsal giant interneurons are excited, while at the same time the ventral giant interneurons (escape behaviour) are inhibited. The dorsal giant interneurons then 'kick' the flight CPG which, in its turn, kicks the wing muscles[12]. After about three wingbeats, the system (wings plus CPG) settles on an eigenrhythm which is passed back to the dorsal giant interneurons, and from there to the sensors, which acquire the overall rhythm and a new threshold[13].

Another extensively studied invertebrate preparation is the lobster stomatogastric ganglion (Selverston et al., 1983) which turned out to be much more flexible than originally expected (Marder, 1984).

Being a relatively simple vertebrate, the lamprey enjoys a rather special status in CPG-research. The lamprey swims by producing sinusoid movements, travelling down its body. The search for the building blocks of the swimming CPG spans already more than a decade. Although several general schemes have been proposed (e.g., Buchanan & Cohen, 1982; Grillner et al., 1988), no definitive scheme is likely to be presented: The number of cells is simply too large to be faithfully modelled. On the other hand, this obviously poses no major problem to the animal. We may need still another family of models to understand how that can be the case.

---

[12] This second 'kick' is our own interpretation. If, however, the CPG were thought to pre-specify the rhythm of the wings, these would not be allowed to have an eigenfrequency. But they have.

[13] Several loops are involved in this process, each one taking an integer number of flight cycle times. From a physical point of view, this is a necessity: There would be no resonance in the system otherwise; From a neurophysiological point of view, however, it remains to be established how loop-time can depend on overall characteristics of network activity.

In higher vertebrates, the situation is even fuzzier. And again, one wonders how it is that they are able to functionally move without much ado.

We summarise with stepping stone 6:

There exist isolated chunks of nervous tissue which, under the appropriate conditions, produce patterned output, similar to the activity one would expect in the intact animal. Models of distributed control may have captured the general function of invertebrate networks. The functional role of higher vertebrate networks, however, has remained elusive so far.

7 Losing overview

In 1969, Kauffman showed that any re-entrant network with n nodes, and neither too few nor too many connections between them, will, upon being stimulated, oscillate in 1 out of $\sqrt{n}$ possible modes (rather than 2^n, as would be theoretically possible): Such networks simplify their own behaviour. In 1973, Levins showed that evolving (or learning) networks with too many connections will lose the surplus whenever they function: In order to reduce the immensity of unconstrained search space, systems such as the nervous system form relatively isolated (i.e., not strongly all-to-all connected) compartments. In *coordination*, these will form clusters ('coupling'); In *control*, each single one of these compartments is addressable and affects overall performance.

In their discussion of "new models for motor control", Altman and Kien (1989) view such compartments as "distributed control systems that do not depend on a central command centre for the execution of behavioural outputs", a suggestion they offer for invertebrates as well as vertebrates. To date, a distributed control system has been implemented in, for instance, an insectoid robot (Brooks, 1989) which is capable of adaptive locomotory behaviour in a cluttered environment.

The functional role of distributed motor control appears to be dual. If there exists any mechanism to switch a relevant network on (or off), then that mechanism is of a low-dimensional nature when compared to the high dimensionality of the network. On the other hand, networks are of much lower dimensionality than the locomotor apparatus: No network has ever been claimed to exactly specify when which muscle fibre has to contract how strongly. It is, then, particularly from the point of view of the locomotor apparatus that CPGs may turn out to have program-like properties.

Take human walking as an example. There are about 8 major muscle groups in the sagittal plane of each leg, which amounts to 16 muscle groups in total. If one assumes that each group switches state about 4 times per cycle, then walking consists of a particular order of 64 possible events—that is to say, in terms of major muscle groups in the sagittal plane. A total of 64 events gives a total of 64! theoretically possible sequences, which amounts to roughly 10^{89} according to our calculator. Given that the universe exists about 10^{17} seconds, a baby, born at the first second of the history of the universe, would have needed 10^{72} trials per second if it were to have tried out all possible sequences by now at least once.

So we need mechanisms to reduce the *a priori* search space if we are ever going to functionally move (achieve any goal) at all. Some of these mechanisms will reside in physical laws. Some, we argue, reside in nervous networks—the advantage over physical laws being that networks are changeable. One may want to cry out at being deprived of so much freedom, but without such constraining (Pattee, 1973) no freedom could have emerged[14].

---

[14] In that respect, learning one's mother tongue is similar: the number of utterances one is confronted with in early childhood, is an extremely small subset of the set of all possible ones; this, however, is exactly how one learns to exploit the search space, even to such an extent that humans produce novel utterances all the time.

Remaining the same, as well as being able to change, then, should both be important properties of neural networks. Maybe the appropriate term is 'soft-wired' rather than hard wired (or not wired at all).

The ability to change, however, has turned out to be so overwhelming, that, presently, researchers again wonder how the high dimensionality of motor control search space can be overcome. In 1991, Harris-Warrick and Marder presented an overview of the possibilities for chemical modulation of network activity:

> The enormous diversity of modulatory actions at every level of neural networks poses two important questions: 1. How do networks retain their essential characteristics and continue to operate stably despite all this modulation? 2. Of the many changes induced in a network by a neuromodulator, which are the most important in determining the final motor pattern, and which provide only subtle alterations? (p. 53.)

That the effects of modulation can be dramatic, has been shown rather elegantly by Meyrand et al. (1991). The lobster stomatogastric ganglion mainly consists of an oesophageal, a gastric, and a pyloric network, responsible for initial pumping, then grinding, and finally filtering of food in the foregut (Selverston et al., 1983). Upon depolarisation of the so-called 'pyloric suppressor' neurons, these networks functionally reorganise themselves into a new network, responsible for the swallowing act (Meyrand et al., 1991; Figure 7). The Meyrand et al. (1991) study offers an example of switching to a qualitatively different state. Other research has revealed modulatory effects where one or a few dimensions of the system change quantitatively, leaving the overall topology intact (Harris-Warrick & Marder, 1991; Grillner & Matsushima, 1991).

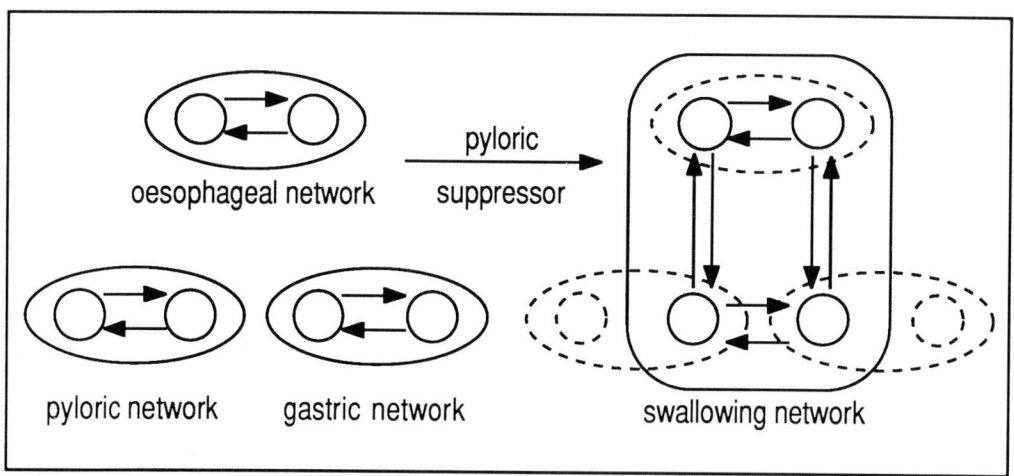

Figure 7. Pyloric suppressor induced network reorganisation in the lobster. (After Meyrand et al., 1991, p. 62.)

Eventually, these modulatory effects may be charted. The present authors expect such charting to take place in terms of coordination: Some low-dimensional description will be found to show how the system simplifies itself. And again, we will run into the problem of control.

90

There is a strange asymmetry in the time-aspects of coordination and control. In coordination, high dimensionality precedes low dimensionality, whereas in control, the order is reversed. In network models, time is kept implicit and the asymmetry remains hidden; in tree models, time is made explicit and the asymmetry is clearly visible (cf. Figure 5).

This symmetry-breaking may be what distinguishes living from non-living systems (Varela, 1979). In physical applications of the theory of dynamical systems, where the microscopy is often homogeneous, it is usually non-essential. In biological applications, however, where the microscopy is always heterogeneous, it is of paramount importance.

From an evolutionary viewpoint, enzymes may offer the first example: They exert immediate influence on the rate of a specific chemical reaction by which they are influenced only indirectly, i.e., with a certain delay. If one is to look at the network of chemical reactions with a very short observation time, everything appears to behave normally (symmetrically). There exists, however, a higher time-scale at which the enzyme can be 'shown' to have acquired 'control', the asymmetry in question being better visualised in a tree than in a network. (Hence, the use of 'cascades' in biochemistry.) An intricate machinery exists to ensure that the enzyme's 'instructions' flow correctly into the high-dimensional space of cellular chemistry.

A similar tale can be told about nervous systems. Parts of these systems sometimes separate themselves out in order to acquire 'control'. An intricate machinery then ensures that their 'instructions' appropriately flow out into the high-dimensional microscopy. The present authors would like to argue that there is an urgent need to analyse such flowing out into high dimensionality. We are aware of two CPG-related examples of this phenomenon: from the low dimensionality of a 'command system' (such as: the cockroach giant dorsal interneuron, Libersat et al., 1989) to the high dimensionality of a CPG; and, from the low dimensionality of a CPG (such as: the coupling schemes for the lamprey spinal cord—Kiemel, 1990) to the high dimensionality of motor unit recruitment.

We have tried to show that the dynamic modelling of cooperative behaviour, is an essential step towards unravelling the system. The essential closure, however, cannot be reached before it is exactly understood how low-dimensional 'instructions' can flow into a high-dimensional work space. So, we don't know what our next stepping stone is going to be.

In living systems, all elements are equal. But there are times when some elements are more equal than other ones.

Acknowledgements. The authors want to gracefully acknowledge Avis Cohen for her fierce stimulation. Both as a hostess and as a critic, she remains unsurpassed. Many students have helped us to lose our overview—among others, Paul Derksen, Jolande Jurrius, Claudine Lamoth, Lieke Peper, and Chris Sybrandy. We are indebted to our colleagues for their willingness to discuss these matters with us—among others, Emilio Bizzi, Dan Bullock, Anatoly Fel'dman, John Fentress, Wim van der Grind, Hermann Haken, Scott Kelso, Nick Mellen, Howard Pattee Hans van Rappard, Robert Rosen, and Arne Wunderlin. Helpful comments to an earlier version of this paper were given by, among others, Avis Cohen, Piet Eikelenboom, Tim Kiemel, Claire Michaels, Lieke Peper, Hans van Rappard, Barry Roberts, and Mark Willems.

References

Adrian, E.P., & Buytendijk, F.J.J. (1931). Potential changes in the isolated brain stem of the goldfish. *The Journal of Physiology, **71**, 121-135.

Altman, J.S., & Kien, J. (1989). New models for motor control. *Neural Computation, **1**, 173-183.

Arnol'd, V.I. (1983). *Geometrical methods in the theory of ordinary differential equations.* Berlin: Springer Verlag.

Ayers, J., Carpenter, G., Currie, S., & Kinch, J. (1983). Which behavior does the lamprey central motor program mediate? *Science*, **221**, 1311-1314.

Babloyantz, A., & Destexhe, A. (1986). Low-dimensional chaos in an instance of epilepsy. *Proceedings of the National Academy of Sciences of the USA*, **83**, 2513-2517.

Beek, P.J. (1989). *Juggling dynamics*. Amsterdam: Free University Press; PhD Thesis.

Beek, P.J., & Meijer, O.G. (1988). On the nature of 'the' motor-action controversy. In: O.G. Meijer, & K. Roth (Eds.), *Complex movement behaviour: 'The' motor-action controversy* (pp. 157-185). Amsterdam: North-Holland.

Bernstein, N.A. (1935). [Das Problem der Wechselbeziehungen zwischen Koordination und Lokalisation.] We used the translation in: L. Pickenhain, & G. Schnabel (1988, Eds.), *Bewegungsphysiologie von N.A. Bernstein* (pp. 67-98). Leipzig: Johann Ambrosius Barth.

Bernstein, N.A. (1940). [Biodynamik der Lokomotionen: Genese, Struktur, Veränderungen.] We used the translation in: L. Pickenhain, & G. Schnabel (1988, Eds.), *Bewegungsphysiologie von N.A. Bernstein* (pp. 21-66). Leipzig: Johann Ambrosius Barth.

Bernstein, N.A. (1957). [Einige heranreifende Probleme der Regulation der Bewegungsakte.] We used the translation in: L. Pickenhain, & G. Schnabel (1988, Eds.), *Bewegungsphysiologie von N.A. Bernstein* (pp. 173-193). Leipzig: Johann Ambrosius Barth.

Brooks, R.A. (1989). A robot that walks: Emergent behaviors from a carefully evolved network. *Neural Computation*, **1**, 253-262.

Brown, T.G. (1914). On the nature of the fundamental activity of the nervous centres; together with an analysis of the conditioning of rhythmic activity in progression, and a theory of the evolution of function in the nervous system. *Journal of Physiology (London)*, **48**, 18-46.

Buchanan, J.T., & Cohen, A.H. (1982). Activities of identified neurons, motoneurons, and muscle fibres during fictive swimming in the lamprey and effects of reticulospinal and dorsal cell stimulation. *Journal of Neurophysiology*, **47**, 948-960.

Calabrese, R. (1980). Invertebrate central pattern generators: Modelling and complexity. *The Behavioral and Brain Sciences*, **3**, 542.

Cohen, A.H. (1980). A new generation of experimental and theoretical methods is needed in neurobiology. *The Behavioral and Brain Sciences*, **3**, 543.

Cohen, A.H., & Wallén, P. (1980). The neuronal correlate of locomotion in fish: "Fictive swimming" induced in an in vitro preparation of the lamprey spinal cord. *Experimental Brain Research*, **41**, 11-18.

Davis, W.J. (1980). Neurophilosophical reflections on central nervous pattern generators. *The Behavioral and Brain Sciences*, **3**, 543-544.

Dennett, D.C. (1979). *Brainstorms: Philosophical essays on mind and psychology*. Hassocks: Harverster Press.

Engberg, I., & Lundberg, A. (1969). An electromyographic analysis of muscular activity in the hindlimb of the cat during unrestrained locomotion. *Acta Physiologica Scandinavica*, **75**, 614-630.

Ermentrout, G.B., & Kopell, N. (1990). Oscillator death in systems of coupled neural oscillators. *Society for Industrial and Applied Mathematics Journal of Applied Mathematics*, **50**, 125-146.

Fukson, O.I., Berkinblitt, M.B., & Fel'dman, A.G. (1980). The spinal frog takes into account the scheme of its body during the wiping reflex. *Science*, **209**, 1261-1263.

Getting, P. (1988). Comparative analysis of invertebrate central pattern generators. In: A.H. Cohen, S. Rossignol, & S. Grillner (Eds.), *Neural control of rhythmic Movements in vertebrates* (pp. 101-128). New York: John Wiley & Sons.

Getting, P., & Dekin, M.S. (1985). Mechanisms of pattern generation underlying swimming in *Tritonia*, IV: Gating of a central pattern generator. *Journal of Neurophysiology*, **53**, 466-479.

Grillner, S. (1985). Neurobiological bases of rhythmic movement acts in vertebrates. *Science*, **228**, 143-149.

Grillner, S., Buchanan, J.T., & Lansner, A. (1988). Simulation of the segmental burst generating network for locomotion in lamprey. *Neuroscience Letters*, **89**, 31-55.

Grillner, S., & Matsushima, T. (1991). The neural network underlying locomotion in lamprey—synaptic and cellular mechanisms. *Neuron*, **7**, 1-15.

Grillner, S., & Zangger, P. (1975). How detailed is the central pattern generation for locomotion? *Brain Research*, **88**, 367-371.

Grillner, S., & Zangger, P. (1979). On the central generation of locomotion in the low spinal cat. *Experimental Brain Research*, **34**, 241-262.

Haken, H. (1983). *Advanced synergetics: Instability hierarchies of self-organising systems and devices*. Berlin: Springer Verlag.

Haken, H., Kelso, J.A.S., & Bunz, H. (1985). A theoretical model of phase-transitions in human hand movements. *Biological Cybernetics*, **51**, 347-356.

Haken, H. & Wunderlin, A. (1990). Synergetics and its paradigm of self-organisation in biological systems. In: H.T.A. Whiting, O.G. Meijer, & P.C.W. van Wieringen (Eds.), *The natural-physical approach to movement control* (pp. 1-36). Amsterdam: VU University Press.

92

Harris-Warrick, R.M., & Marder, E. (1991). Modulation of neural networks for behavior. *Annual Review of Neuroscience*, **14**, 39-57.

Hinton, G. (1984). Some computational solutions to Bernstein's problems. In H.T.A. Whiting (Ed.), *Bernstein reassessed: Human motor actions* (pp. 413-440). Amsterdam: North-Holland.

Jantsch, E. (1985). *The self-organizing universe: Scientific and human implications of the emerging paradigm of evolution*. Oxford: Pergamon Press.

Kauffman, S.A. (1969). Metabolic stability and epigenesis in a randomly constructed genetic net. *Journal of Theoretical Biology*, **22**, 437-467.

Kiemel, T.L. (1990). *Three problems from the mathematics of neural oscillations*. Bethesda, MD: NIH; PhD Thesis.

Kugler, P.N., & Turvey, M.T. (1987). *Information, natural law, and the self-assembly of rhythmic movement*. Hillsdale, NJ: Lawrence Erlbaum.

Kupfermann, I., & Weiss, K.R. (1978). The command-neuron concept. *The Behavioral and Brain Sciences*, **1**, 3-39.

Levins, R. (1973). The limits of complexity. In: H.H. Pattee (Ed.), *Hierarchy Theory: The challenge of complex systems* (pp. 109-127). New York: George Braziller.

Libersat, F., & Camhi, J.M. (1988). Control of cercal position during flight in the cockroach: A mechanism for regulating sensory feedback. *Journal of Experimental Biology*, **136**, 483-488.

Libersat, F., Levy, A., & Camhi, J.M. (1989). Multiple feedback loops in the flying cockroach: Excitation of the dorsal and inhibition of the ventral giant interneurons. *Journal of Comparative Physiology A*, **165**, 651-668.

Lotze, R.H. (1853). [Recension von Eduard Pflüger, die sensorischen Functionen des Rückenmarks der Wirbelthiere nebst einer neuen Lehre über die Leitungsgezetze der Reflexionen.] We used the reprint in: D. Pfeiffer (1891, Ed.), *Kleine Schriften von Hermann Lotze, Band 3* (pp. 191-209). Leipzig: S. Hirzel.

Marder, E. (1984). Mechanisms underlying neurotransmitter modulation of a neuronal circuit. *Trends in Neurosciences*, **7**, 48-53.

McDougall, W. (1903). [The nature of inhibitory processes within the nervous system. *Brain*, **26**, 153-191.] Reference taken from: Friesen, 1980.

Meijer, O.G., & Bongaardt, R. (1992). Synergetics, self-simplification, and the ability to undo. In: R. Friedrich & A. Wunderling (Eds.), *Evolution of dynamical structures in complex systems. Springer Proceedings in Physics, Vol. 69* (pp. 272-298). Heidelberg: Springer Verlag.

Meijer, O.G., & Roth, K. (1988). *Complex movement behaviour: 'The' motor-action controversy*. Amsterdam: North-Holland.

Meijer, O.G., Wagenaar, R.C. ,& Blankendaal, F.C.M. (1988). The hierarchy debate: Tema con variazioni. In: O.G. Meijer, & K. Roth (Eds.), *Complex movement behaviour; 'The' motor-action controversy* (pp. 489-561). Amsterdam: North-Holland.

Meyrand, P., Simmers, J., & Moulins, M. (1991). Construction of a pattern-generating circuit with neurons of different networks. *Nature*, **351**, 60-63.

Pattee, H.H. (1973). The physical basis and origin of hierarchical control. In: H.H. Pattee (Ed.), *Hierarchy theory: The challenge of complex systems*. New York: George Braziller.

Pattee, H.H. (1977). Dynamic and linguistic modes of complex systems. *International Journal of General Systems*, **3**, 259-266.

Pflüger, E.F.W. (1853). *Die sensorischen Functionen des Rückenmarks der Wirbelthiere nebst einer neuen Lehre über die Leitungsgesetze der Reflexionen*. Berlin: August Hirschwald.

Prigogine, I., & Stengers, I. (1984). [*Order out of Chaos*.] We used the 1986 Fontana edition (London).

Rand, H., Cohen, A.H., & Holmes, P.J. (1988). Systems of coupled oscillators as models of central pattern generators. In: A.H. Cohen, S. Rossignol, & S. Grillner (Eds.), *Neural control of rhythmic movement in vertebrates* (pp. 333-368). New York: John Wiley & Sons.

Rosen, R. (1974). Planning, management, politics and strategies: Four fuzzy concepts. *International Journal of General Systems*, **1**, 245-252.

Rosen, R. (1985). *Anticipatory systems: Philosophical, mathematical, and methodological foundations*. Oxford: Pergamon Press.

Rosen, R. (1991). *Life itself: A comprehensive inquiry into the nature, origin, and fabrication of life*. New York: Columbia University Press.

Russell, B. (1979). *Introduction to mathematical philosophy*. New York: MacMillan.

Selverston, A.I. (1980). Are central pattern generators understandable? *The Behavioral and Brain Sciences*, **3**, 535-571.

Selverston, A.I., Miller, J.P., & Wadepuhl, M. (1983). Neuronal mechanisms for rhythmic motor pattern generation in a simple system. In: R. Herman, S. Grillner, P. Stein, & D. Stuart (Eds.), *Neural control of locomotion* (pp. 377-399). New York: Plenum.

93

Van Soest, A.J. (1992). *Jumping from structure to control: A simulation study of explosive movements.* Amsterdam: Thesis Publishers; PhD Thesis.

Varela, F.J. (1979). *Principles of biological autonomy.* New York: North-Holland.

Verworn, M. (1907). *Physiologisches Praktikum für Mediziner.* Jena: Gustav Fischer.

Von Holst, E. (1937). [Vom Wesen der Ordnung im Zentralnervensystem. *Naturwissenschaften, 25,* 625-647.] We used the translation in: Martin, R. (1973, Ed.). *The behavioural physiology of animals and man: The collected papers of Erich von Holst, Volume 1* (pp. 3-32). Coral Gables, FL: University of Miami Press.

Wallén, P. & Williams, T.L. (1984). Fictive locomotion in the lamprey spinal cord *in vitro* compared with swimming in the intact and spinal animal. *Journal of Physiology (London),* **347,** 225-239.

Williams, T.L., Sigvardt, K.A., Kopell, N., Ermentrout, G.B., & Remler, M.P. (1990). Forcing of coupled nonlinear oscillators: Studies of intersegmental coordination in the lamprey locomotor central pattern generator. *Journal of Neurophysiology,* **64,** 862-871.

Wilson, D.M. (1961). The central nervous control of flight in a locust. *Journal of Experimental Biology,* **38,** 862-871.

Author Index